Robin J. Green

Editor

Viral Infections in Children, Volume I

Springer

Editor
Robin J. Green
Department of Paediatrics and Child Health
University of Pretoria, School of Medicine
Pretoria, ZA
South Africa

ISBN 978-3-319-54032-0   ISBN 978-3-319-54033-7   (eBook)
DOI 10.1007/978-3-319-54033-7

Library of Congress Control Number: 2017939124

Printed on acid-free paper

This Springer imprint is published by Springer Nature
The registered company is Springer International Publishing AG
The registered company address is: Gewerbestrasse 11, 6330 Cham, Switzerland

This book is dedicated to the many children who are sick with a viral infection. May I also dedicate it to my partner and best friend—Jessica.

Preface

This book is important. It is important not only because some of the world's leading authorities have summarized the evidence on each of these topics but also important because we live in a world where illness and infection are often perceived to be due to bacteria. This has led to an enormous increase in antibiotic use and, of course, the attendant increases in antimicrobial resistance. We need to know, and make our colleagues aware, of the enormous burden of viral infections in children, for which conventional antibiotics are redundant and unnecessary and even harmful. Knowing the burden may now save our world from the looming threat of "superbugs". We do of course need to work on dealing with the additional threat from viruses and much research and effective therapy is still needed. Hopefully some of us reading this book will be inspired to provide these outcomes.

This book attempts to summarize the known literature on each of the organ systems where viruses may cause disease. In addition, there are chapters on viral illnesses where the organisms produce more profound systemic pathologies. Since this book is inclusive, I have added a chapter on HIV and the problems caused specifically by the virus itself and another on the comorbid illnesses seen due to compromised immunity.

Each of the chapters contains the latest information on the pathobiology of viral infections, the clinical presentation and diagnostic pointers as well as management strategies. In this sense, the book should appeal to both non-clinicians and clinical specialists alike. I am particularly hoping that young doctors will find the book valuable in treating patients, because after all this should be our mandate—helping children in distress.

One word of explanation. Those of you who read the whole book may notice that some conditions and concepts are repeated in different chapters. This is for a number of reasons including emphasis on a different aspect of the disease, that repeating a certain topic may be required in order to introduce other concepts in that chapter, that the disease is common or currently making headlines and because the book is available on-line and some readers may elect to "pick-up" only selected chapters.

Enjoy the read!

Pretoria, ZA, South Africa   Robin J. Green

Contents

Congenital Viral Infections

1

Gamal Samy Aly, Hesham Abdel-Hady,
and Maged Z. Ibrahim

Abstract

Infections acquired in-utero or during the birth process are a significant cause of fetal and neonatal mortality and morbidity, sometimes effects being delayed into late childhood. The newborn infant with a congenital infection may be completely asymptomatic or demonstrate abnormal/restricted growth, poor weight gain, developmental delay, or multiple laboratory abnormalities. The "TORCH" acronym is well recognized in the field of neonatal/perinatal medicine. However, there are other well-described causes of in-utero infection, including varicella zoster virus, and parvovirus B19. The chapter will give a concise, yet informative, description of the most commonly observed congenital viral infections. Other congenital infections are beyond the scope of this title.

1.1 Congenital Zika Virus Infection

Zika virus (ZIKV) is an emerging mosquito-borne flavivirus whose incidence and prevalence has exploded in 2015, in the Americas. Investigations reveal in-utero transmission of ZIKV to the fetus in pregnant women. These vertical infections have been associated with the potential for serious outcomes including, but not limited to,

G.S. Aly, M.D. (✉) • M.Z. Ibrahim, M.B.B.Ch., M.Sc.
Ain Shams University, Cairo, Egypt
e-mail: drgamalsamy@gmail.com; maged.zakaria@yahoo.com;
maged.zakaria@chi.asu.edu.eg

H. Abdel-Hady, M.D.
Neonatal Intensive Care Unit, Mansoura University Children's Hospital, Mansoura, Egypt
e-mail: hehady@yahoo.com

© Springer International Publishing AG 2017
R.J. Green (ed.), *Viral Infections in Children, Volume I*,
DOI 10.1007/978-3-319-54033-7_1

microcephaly and spontaneous abortions. Because of these epidemiological and clinical features, the World Health Organization (WHO) declared ZIKV disease a "Public Health Emergency of International Concern" under the International Health Regulations of 2005 on February 1, 2016.

1.1.1 Vectors

Humans and non-human primates are the likely principal vertebrate hosts for ZIKV, which is primarily transmitted to humans through the bite of mosquitoes, most commonly *Aedes aegypti* and possibly *Aedes albopictus*. *Aedes* species mosquitoes are aggressive daytime feeders. They live in and around human households, are difficult to eradicate, and are able to reproduce in small water containers [1].

1.1.2 Routes of Transmission

Mosquito-borne transmission is the main route of maternal acquisition of ZIKV. Intrauterine transmission has been documented, resulting in congenital infection. Neonatal infection, therefore, results from intrapartum transmission from a viremic mother to her newborn [2, 3].

ZIKV RNA can be present in breast milk. However, based on current evidence, the potential risk of ZIKV transmission through breast milk is outweighed by the known benefits of breastfeeding [4]. Similarly, no evidence exists that ZIKV can be transmitted through saliva or urine, although ZIKV RNA has been found in these fluids [5]. Infection has been reported to occur via laboratory blood exposure and sexual routes [6, 7].

1.1.3 Pathogenesis

Studies in animals and human placental studies support the hypothesis that maternal infection leads to placental infection and injury, followed by transmission of the virus across the placenta and ultimately to the fetal brain, where it targets and destroys neuronal progenitor cells and, to a lesser extent, neuronal cells at all stages of maturity [8, 9]. Neuronal growth, proliferation, migration, and differentiation are disrupted, thus slowing and impairing normal brain development.

1.1.4 Clinical Features

Although maternal–fetal transmission of ZIKV during pregnancy has been documented, the incidence of congenital ZIKV infection and the frequency of adverse outcomes among pregnancies infected with ZIKV, are unknown [2, 3].

Microcephaly is the adverse infant outcome for which there is the most evidence. Between March 2015 and April 2016, more than 5000 cases of microcephaly were reported among newborns born to Brazilian mothers with ZIKV infection; this represents a >20-fold increase in microcephaly compared with previous years. However, no standard definition for microcephaly exists as studies have used different cutoffs (>2 or 3 SDs below the mean for gestational age and gender or below the third or fifth percentile). This problem has led to discussion as to whether or not this increase might be due to misdiagnosis related to different cutoffs or over-reporting related to increased awareness of the possible association with ZIKV [10].

Other congenital anomalies identified on fetal ultrasound and MRI included brain atrophy and asymmetry, hydranencephaly, ventriculomegaly, cerebral calcifications, abnormally formed or absent brain structures (e.g., corpus callosum, thalami, pons, cerebellar vermis, brainstem), bilateral cataracts, intraocular calcifications, and hydrops fetalis [2, 3].

In addition to microcephaly, postnatal neurologic examination findings may include arthrogryposis [11], hypertonia, dysphagia as well as seizures [2, 3]. These findings have led researchers to suggest that ZIKV is neurotropic causing disruption of fetal brain growth that leads to skull collapse [12, 13]. ZIKV might also cause placental damage. Other system and organ involvement, other than the brain, cannot be excluded from current data.

Based on microcephaly due to other causes, infants with severe microcephaly associated with ZIKV infection are likely to be at risk for long-term adverse outcomes, including seizures, cognitive impairment, and hearing and vision losses [14]. No long term data on the outcome of infants with microcephaly related to ZIKV infection is available at this time [2, 3]. It is also unknown whether other central nervous system manifestations beyond microcephaly might occur (e.g. cognitive impairment without microcephaly).

Eye abnormalities (including cataracts, microphthalmia, chorioretinal atrophy, optic nerve hypoplasia and pallor, and lens subluxation) have been described in infants born with microcephaly, and who are suspected of having congenital ZIKV infection [15].

1.1.5 Diagnosis

1.1.5.1 Indications for Laboratory Testing

Based on current CDC recommendations (June 2016), infants born to mothers who traveled to, or resided in areas with local ZIKV transmission (Fig. 1.1) during pregnancy, or who are born to mothers who had sexual contact with male partners who traveled to, or resided in, these areas, might be at risk for congenital ZIKV infection. All infants with the aforementioned epidemiologic risk factors and microcephaly (defined by [2, 3] as occipitofrontal circumference at birth less than the third percentile for gestational age and sex and not explained by other etiologies) or intracranial calcifications should be tested for ZIKV, regardless of maternal test results.

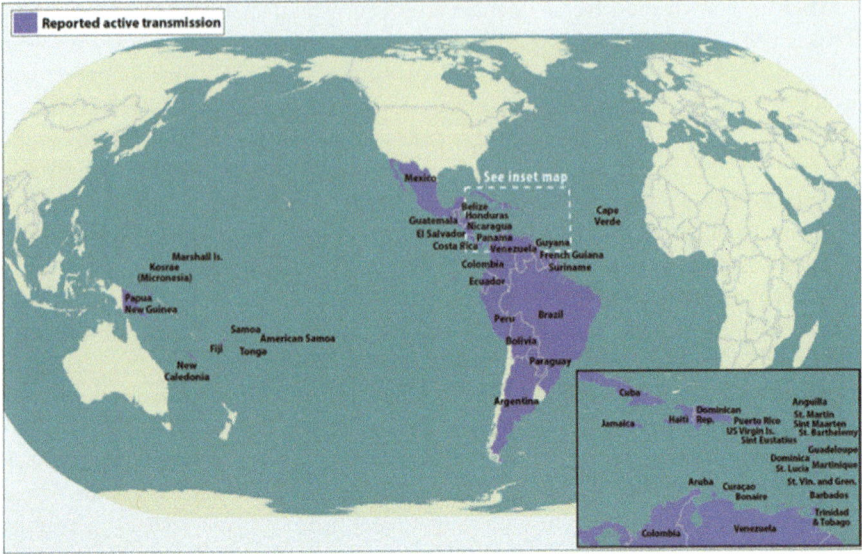

Fig. 1.1 Countries reporting active mosquito transmission of ZIKV (source: CDC travel information, 2016)

1.1.5.2 Laboratory Testing Options

Reverse transcription-polymerase chain reaction (RT-PCR) testing of amniotic fluid, placenta, fetal serum, fetal brain tissue, and fetal cerebrospinal fluid is highly sensitive and specific for detecting ZIKV RNA [16].

The infant's serum and urine should be also tested for ZIKV immunoglobulin M (IgM) and neutralizing antibodies, as well as dengue virus IgM and neutralizing antibodies to discriminate between cross-reacting antibodies. The initial sample should, if possible, be collected from the umbilical cord or directly from the infant within 2 days of birth. If infant cerebrospinal fluid (CSF) is available, the CSF should also be tested for ZIKV RNA (via RT-PCR), ZIKV IgM and neutralizing antibodies, and dengue virus IgM and neutralizing antibodies to discriminate between cross-reacting antibodies. CSF specimens need not be collected for the sole purpose of ZIKV testing but may be reasonable for evaluation of infants with microcephaly or intracranial calcifications [2, 3].

Histopathologic examination of the placenta and umbilical cord should be considered, with ZIKV immunohistochemical staining on fixed tissue and ZIKV RNA (via RT-PCR) on fixed and frozen tissue [6, 7]. WHO criteria for congenital ZIKV case classification are listed in Table 1.1.

Because of the aforementioned cross-reactivity in patients with previous flavivirus exposure through natural infection or immunization (e.g., yellow fever vaccine), health care providers should consult with public health authorities for assistance in interpreting test results [2, 3].

Table 1.1 Congenital ZIKV case classification

Probable
An infant meets the clinical criteria (microcephaly or intracranial calcifications or central nervous system abnormalities) **AND**:
• Mother lived in or traveled to a country or area with ongoing ZIKV transmission during the pregnancy; **OR**
• Mother has laboratory evidence of ZIKV or unspecified flavivirus infection during pregnancy
AND the infant meets the following laboratory criteria:
• ZIKV IgM antibodies detected in serum or CSF; **AND**
• Tests negative for dengue or other endemic flavivirus-specific IgM antibodies; **AND**
– No neutralizing antibody testing performed; **OR**
– Less than fourfold difference in neutralizing antibody titers between ZIKV and dengue or other flaviviruses endemic to the region where exposure occurred
Confirmed
An infant meets the clinical criteria **AND** meets one of the following laboratory criteria:
• ZIKV detection by culture, antigen test, or polymerase chain reaction (PCR) in serum, CSF, amniotic fluid, urine, placenta, umbilical cord, or fetal tissue; **OR**
• ZIKV IgM antibodies present in serum or CSF with ZIKV neutralizing antibody titers fourfold or greater than neutralizing antibodies against dengue or other flaviviruses endemic to the region where exposure occurred

Source: [17]

1.1.6 Differential Diagnosis

Vertical transmission of other flaviviruses seems to occur rarely; this transmission has not been associated with an increased risk for congenital anomalies [18].

Maternal–fetal transmission of other non-flavivirus infections (e.g., rubella virus, cytomegalovirus, lymphocytic choriomeningitis virus, *Toxoplasma gondii*) has been associated with microcephaly. Other manifestations of these congenital infections include brain abnormalities (e.g., intracranial calcifications, hydrocephalus), eye abnormalities (e.g., cataracts, glaucoma, chorioretinitis), and hearing impairment [1].

Congenital microcephaly can be associated with prenatal exposure to other infectious and noninfectious (e.g., alcohol, mercury) agents and with genetic conditions [14]; evaluation for other etiologies needs to be completed when congenital ZIKV infection is excluded.

1.1.7 Treatment

No specific treatment of congenital ZIKV infection is currently available [1]. Supportive measures should address specific medical and neurodevelopmental issues for the infant's particular needs; research is ongoing to better understand what services will be most effective for these children as they grow. The recommended

Table 1.2 Recommended evaluation and long-term follow up for infants with possible congenital ZIKV infection

Evaluation for all infants with positive or inconclusive ZIKV test results
• Thorough physical examination, including careful measurement of head circumference, length, weight, and assessment of gestational age
• Evaluation for neurologic abnormalities, dysmorphic features, enlarged liver or spleen, and rash/other skin lesions (including photographs of any rash, skin lesions, or dysmorphic features)
• Cranial ultrasound (unless performed in the third trimester and clearly showed no brain abnormalities)
• Ophthalmologic evaluation (including retina) before hospital discharge or within 1 month after birth
• Evaluation of hearing by evoked otoacoustic emissions testing or auditory brainstem response testing before hospital discharge or within 1 month after birth
• Consultation with appropriate specialist for any abnormal findings
Additional evaluation for infants with positive or inconclusive ZIKV test results who have microcephaly or other findings consistent with congenital ZIKV infection
• Consultation with clinical geneticist or dysmorphologist
• Consultation with pediatric neurologist
• Testing for other congenital infections (such as cytomegalovirus, toxoplasmosis, and rubella); consider consultation with pediatric infectious disease specialist
• Complete blood count, platelet count, and liver function and enzyme tests
• Genetic or other teratogenic causes should be considered if additional anomalies are identified through history, clinical examination, or imaging studies
Long-term follow-up for infants with positive or inconclusive ZIKV test results
• Additional hearing screen at 6 months of age and audiology follow up of abnormal newborn hearing screening
• Continued evaluation of developmental characteristics and milestones, as well as head circumference, through first year of life
• Consultation with appropriate medical specialists (e.g., pediatric neurology, developmental and behavioral pediatrics, physical and speech therapy) if any abnormalities are noted and as concerns arise

Source: [2, 3]

evaluation and long-term follow up for infants with possible congenital ZIKV infection is summarized in Table 1.2.

Mothers are encouraged to breastfeed infants even in areas where ZIKV is found, as available evidence indicates the benefits of breastfeeding outweigh any theoretical risks associated with ZIKV infection transmission through breast milk [19].

1.1.8 Prevention

No vaccine is available to prevent ZIKV infection [2, 3]. Males who reside in or have traveled to an area of active ZIKV transmission and have a pregnant partner should abstain from sexual activity or consistently and correctly use condoms during sex for the pregnancy duration [2, 3].

The US Food and Drug Administration issued guidance on February, 2016, regarding deferral of blood donations from persons who have traveled to areas with active ZIKV transmission, have potential exposure to the virus, or have had a confirmed ZIKV infection [20].

Based on Center for Disease Control (CDC) interim travel guidance, pregnant women should postpone travel to areas with ongoing ZIKV transmission. Pregnant women who must travel to one of these areas should strictly follow steps to prevent mosquito bites during their trip. Personal avoidance measures include staying in buildings with air-conditioning or with window and door screens, wearing full length garments and socks, and using mosquito repellent. Bed nets are recommended where accommodations are not adequately screened or air-conditioned [2, 3, 21].

The CDC also recommends the use of insect repellents registered by the Environmental Protection Agency according to the instructions on the label [2, 3]. Products containing N,N-Diethyl-meta-toluamide (DEET) should not be used on children aged <2 months; this group can be protected by use of mosquito netting. Mosquito repellents containing oil of lemon eucalyptus (p-Menthane-3,8-diol) should not be used in children aged <3 years. Only adults should handle repellents, which should be applied judiciously to children's exposed skin, avoiding the hands, eyes, mouth, and broken or irritated skin. Skin treated with mosquito repellent should be washed with soap and water after returning indoors, especially before meals. Combination products that include both mosquito repellent and sunscreen should be avoided because sunscreen may need to be applied more frequently and in larger amounts than needed for adequate mosquito bite protection. Sunscreen, when used, should be applied before repellent [1].

1.2 Congenital Rubella Infection

The term congenital rubella infection (CRI) refers to all outcomes associated with intrauterine rubella infection (e.g., miscarriage, stillbirth, combinations of birth defects, asymptomatic infection). While congenital rubella syndrome (CRS) includes variable patterns of birth defects (e.g., hearing defect, congenital heart diseases, cataracts/congenital glaucoma, pigmentary retinopathy) [22].

Rubella infection and the CRS have largely been eliminated in the United States [23]. However, rubella outbreaks continue to occur in other parts of the world. Immigrants have been the source of CRS in some developed countries, such as the United Kingdom [24] and Canada [25].

1.2.1 Routes of Transmission and Pathogenesis

The human-specific RNA rubella virus is a member of the Togavirus family. Humans are the only known hosts, with an incubation period of approximately 18 days after contact [26]. Maternal rubella virus infection occurs through respiratory secretions and also from stool, urine, and cervical secretions. Transplacental fetal infection can occur at any time during pregnancy, but early-gestation infection may result in multiple organ anomalies [27].

1.2.2   Pathogenesis

Maternal-fetal transmission of rubella virus occurs via hematogenous spread during maternal viremia, which usually occurs 5–7 days after maternal infection. After infecting the placenta, the virus spreads through the vascular system of the developing fetus. The resulting congenital defects stem from cytopathic damage to blood vessels and ischemia in affected organs [28]. Fetal infection is chronic, persisting throughout gestation and after birth.

The risk of maternal-fetal transmission varies depending upon the time of maternal infection, with the highest risk in the first 10 weeks of gestation. The clinical manifestations also vary depending upon the time of maternal infection. Structural cardiac and eye defects typically result when maternal infection occurs before 8 weeks, whereas hearing loss may be observed in maternal infections until 18 weeks gestation. No abnormalities were found when fetal infection occurred beyond the 20th week of gestation [29]. Fetal growth restriction may be the only sequela of third-trimester infection [30].

There are two proposed mechanisms for rubella cytopathology: virus-induced inhibition of cell division and direct cytopathic effects. Support for virus-induced inhibition of cell division is provided by observations that organs of congenitally infected infants are smaller and contain fewer cells than those of uninfected infants [31], mitotic activity is depressed in congenitally infected embryonic primary cell cultures [32], cell division is slowed in human fetal cells that have been infected with rubella virus in vitro [33], and a protein extracted from rubella virus-infected human fetal cells inhibits mitosis in uninfected cells [34].

Support for direct cytopathic effects is provided by an *in-vitro* study demonstrating variable degrees of rubella virus-induced programmed cell death (apoptosis) in different cell lines [35]. These findings suggest that programmed cell death depends upon unique cellular properties and may provide an explanation for selective organ damage in CRS. Subsequent studies demonstrating cell-specific rubella virus-induced apoptosis (in chorionic villi explants, monolayers of cytotrophoblasts, and adult lung fibroblasts, but not in primary human fetal fibroblasts) may help to explain the persistence of virus in CRS [36]. Other possible explanations for the persistence of rubella virus in CRS include defects in cell-mediated immunity and selective immune tolerance to the rubella virus E1 protein [37].

1.2.3   Clinical Features

The incubation period for rubella is 12–23 days. The infectious period is from 7 days before to 5–7 days after rash onset [38]. Although rubella is asymptomatic in 25–50% of cases, some individuals may experience mild prodromal symptoms such as low-grade fever, conjunctivitis, sore throat, coryza, headaches or malaise, and tender lymphadenopathy for 1–5 days before the onset of the scarletiniform rash, which may be mildly pruritic [39]. The rash characteristically begins on the face and spreads to the trunk and extremities. It will usually resolve within 3 days in the same order in which it appeared (face first and then body) [40]. Polyarthritis and polyarthralgia are potential sequelae about 1 week after the rash [41].

Spontaneous abortion is seen in 20% of cases when rubella occurs in the first 8 weeks of pregnancy. Neonatal presentation includes low birth weight, meningoencephalitis, jaundice, hepatosplenomegaly, hemolytic anemia, interstitial pneumonia, and thrombocytopenia with or without purpura (Fig. 1.2: blueberry muffin lesions) [42].

Approximately one-half of children infected during the first 2 months of gestation have congenital heart disease. Patent ductus arteriosus and pulmonary artery branch stenosis (Fig. 1.3) are the most common lesions [43]. Ophthalmic permanent manifestations in the neonate include microphthalmos, cataracts (25%) (Fig. 1.4),

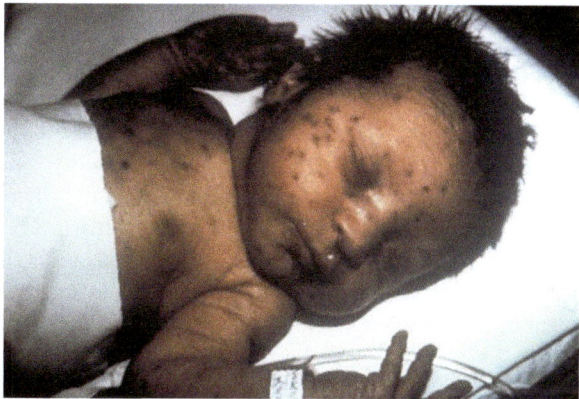

Fig. 1.2 "Blueberry muffin" skin lesions in a case of congenital rubella syndrome (Source: Centers for Disease Control and Prevention Public Health Image Library)

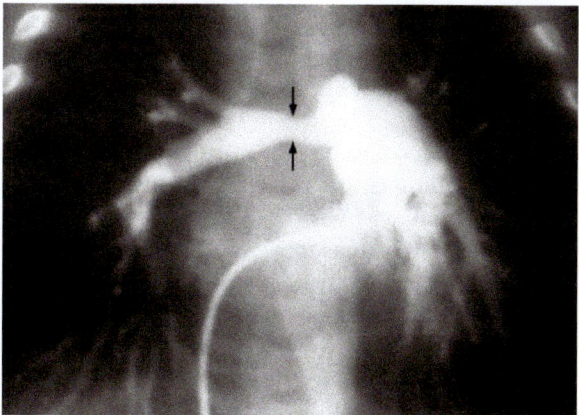

Fig. 1.3 An angiocardiogram highlighting the pulmonary arterial tree. It revealed the presence of a stenosis of the right pulmonary artery in a case of congenital rubella syndrome (Source: Centers for Disease Control and Prevention Public Health Image Library)

Fig. 1.4 Cataracts due to
congenital rubella
syndrome (Source:
Centers for Disease
Control and Prevention
Public Health Image
Library)

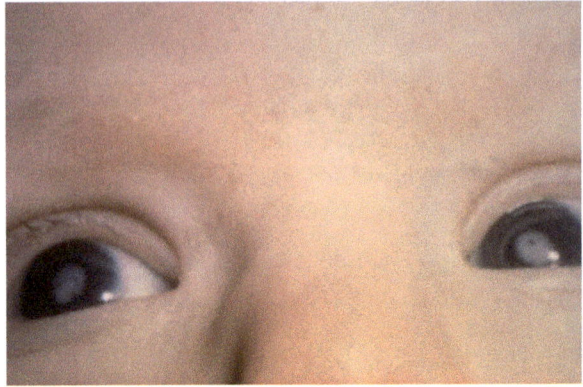

cloudy cornea, infantile glaucoma, iris hypoplasia, and "salt and pepper" retinopathy (with minimal effect on vision) [44].

Microcephaly, and unilateral or bilateral sensorineural or central deafness are often present (nerve deafness is the single most common finding among infants with CRS). Motor delay, behavioral disorders, autism, and psychiatric disorders are less common [42].

It should be noted that CRS is a progressive disease due to the persistence of the viral infection and the defective immune response to the virus. Late onset manifestations can include deafness, intellectual disability, speech defects/language delay, behavioral and educational difficulties, type 1 diabetes mellitus (20%), thyroid disorders, and progressive rubella panencephalitis (PRP). Patients with CRS have a high prevalence of pancreatic islet cell surface antibodies and antithyroid antibodies, suggesting that immune-mediated mechanisms may play a role in the development of diabetes and thyroid disease [45].

1.2.4 Diagnosis

The diagnosis of acute maternal rubella in pregnancy requires serologic testing, best performed within 7–10 days after the onset of the rash and should be repeated 2–3 weeks later [46]. Antenatal diagnosis of fetal infection is performed through detection of rubella-specific IgM in fetal blood by percutaneous umbilical cord blood sampling (PUBS). Postnatally, rubella-specific IgM in neonatal blood, or persistent rubella-specific IgG titers over time, can be detected. Virus can be isolated from urine and oropharynx specimens [47].

Non-specific findings in neonatal investigations can include thrombocytopenia, metaphyseal radiolucencies (osteoporosis) in long-bone films. CSF examination may reveal encephalitis with high protein and cell count. Congenital heart disease is detected by echocardiography. Ophthalmologic and audiologic examinations are always recommended. All newborns who fail hearing screening should undergo evaluation for possible congenital infections (including rubella) [42].

1.2.5 Treatment

There is no specific treatment for CRS. Most neonates with CRS have multiple problems and warrant multidisciplinary treatment. They may require medical, surgical, developmental, and rehabilitative interventions [42]. Although purpura and petechiae may be severe, clinically significant bleeding is uncommon. Management should focus on the associated malformations. Ophthalmologic problems (i.e. cataracts, glaucoma, strabismus) are managed in the same manner as they are in children without CRS. Early intervention for hearing loss is of special importance, as it may improve long-term outcomes in children with CRS. Long-term follow-up is needed to manage late-onset manifestations. A small number of patients with CRS have low levels of serum immunoglobulin G (IgG). Serum immunoglobulins should be measured when clinically indicated (e.g. in children with recurrent infections) [26].

Women with acute infection should be offered pregnancy termination, especially prior to 16 weeks' gestation. Instances of CRS between 16 and 20 weeks' gestation are rare (<1%) and may be manifested by sensorineural deafness (often severe) in the newborn. Appropriate counseling to the non-immune pregnant woman should be provided [46]. No studies have documented CRS after 20 weeks [22].

1.2.6 Prevention

The live attenuated measles, mumps, rubella (MMR) vaccine should not be administered to women known to be pregnant or attempting to become pregnant. Because of the theoretical risk to the fetus when the mother receives a live virus vaccine, women should be counseled to avoid becoming pregnant for 28 days after receipt of MMR vaccine [48]. However, there has been no report of CRS in the offspring of women inadvertently vaccinated during early pregnancy [49]. Rubella vaccine virus has been isolated from breast milk. However, breast-feeding is not a contraindication to vaccination.

The routine use of immunoglobulin for susceptible (non-immune) pregnant women exposed to rubella is not recommended and is considered only if termination of pregnancy is not an option because of maternal preference. Intramuscular immunoglobulin 0.55 mL/kg may be used in such circumstances. Immunoglobulin may decrease clinically apparent infection but does not guarantee prevention of fetal infection [50].

Children with CRS are considered to be contagious until at least 1 year of age unless two cultures of specimens obtained 1 month apart are negative for rubella virus after 3 months of age [51]. Droplet precautions should be instituted for hospitalized patients as soon as CRS is suspected [42]. Children with CRS who remain contagious, should be cared for only by individuals who are immune to rubella (i.e., have received at least one dose of rubella-containing vaccine on or after the first birthday or have a positive serologic test for rubella-specific immunoglobulin G (IgG) antibody) [38].

1.2.7 Prognosis

The risk of mortality is increased in neonates with severe defects (e.g., extreme prematurity, extensive meningoencephalitis, gross cardiac lesions or myocarditis with early heart failure, fulminant interstitial pneumonitis, and rapidly progressive hepatitis) [52]. Long-term follow-up (at 25, 50, and 60 years) of a cohort of 50 patients born during the 1939–1943 rubella epidemic in Australia [53–55] provides the following information about long-term outcome:

- At 25 years, 48 of 50 patients were deaf, 43 of the 48 had severe bilateral deafness, and hearing impairment was detected in all patients by 7 years of age, in most patients by 3 years of age, but in only five patients by 1 year of age
- At 25 years, 11 patients had congenital cardiovascular defects (patent ductus arteriosus, pulmonary stenosis, elevated systemic blood pressure); only 2 of these were detected in the first year of life. At 60 years, 21 of 32 patients had mild aortic valve sclerosis on echocardiography, and 12 were being treated for hypertension
- At 25 years, 26 of 50 patients had typical rubella cataracts or chorioretinopathy; at 60 years, virtually all 32 patients had ocular conditions: rubella retinopathy (12), glaucoma (8), cataracts with onset between the 50- and 60-year follow-up (3), blindness (1), other ocular conditions (8)
- At 25 years, 50% of patients were below the tenth percentile for weight and/or height; at 50 years, 6 of 40 patients (15%) were below the third percentile for height
- At 50 years, 5 of 40 patients (12.5%) were diabetic; at 60 years, the prevalence of type 2 diabetes (22%), thyroid disorders (19%), early menopause (73%), and osteoporosis (12.5%) was increased compared with the Australian population.

1.3   Congenital Cytomegalovirus Infection

Cytomegalovirus (CMV) is a deoxynucleic acid (DNA) virus and a member of the Herpesviridae family (human herpesvirus 5) [56]. CMV is the most common cause of congenital infection [57]. The prevalence of CMV in women of childbearing age varies by region and socioeconomic status and ranges from 40 to 83%. Older age, lower socioeconomic and educational status, and non-white ethnicity are all associated with higher rates of CMV infection [58, 59].

1.3.1   Routes of Transmission

Congenital CMV infection may follow primary infection or reactivation during pregnancy. Prior immunity to CMV is not necessarily protective against reinfection, reactivation, or fetal disease [60]. Approximately 1–4% of seronegative women will acquire a primary CMV infection during pregnancy, and most of them will be asymptomatic [61]. The virus can be transmitted to the mother through body secretions, including saliva, tears, semen, urine, cervical secretions, and CMV-positive blood products [62]. Fetal infection occurs in approximately 32% of cases of primary maternal infections. The risk of vertical transmission to the fetus is significantly higher with primary maternal infection than with recurrent infection (32% versus 1.4%) [63]. Infection in early pregnancy causes more severe fetal infection with significant CNS sequelae [64].

In addition to transplacental transmission, CMV may also be transmitted to the infant intrapartum (through exposure to CMV in cervical secretions). Perinatal or postnatal infection is more common than congenital infection but rarely results in symptomatic infection, and does not have the neurologic sequelae seen in congenitally infected infants [65].

1.3.2 Pathogenesis

In congenital CMV infection, the virus is transmitted to the placenta villi via infected leukocytes causing villitis. The virus passes from the trophoblastic layer into the stromal layer and then into the endothelial cells of the fetal blood vessels after which, the virus infects the fetus hematogenously. Once the fetus is infected, viral replication occurs primarily in the tubular epithelium of the kidneys. The virus is then excreted in fetal urine and subsequently in amniotic fluid. The fetus ingests the infected amniotic fluid, the virus replicates further in the oropharynx, enters the fetal circulation, and spreads to target organs [66]. The pathogenesis of organ damage in congenital CMV infection is not only caused by direct viral cytopathic damage to the fetus, but also from indirect effects through placental dysfunction [67]. The pathogenesis of CMV-related CNS damage is unknown. Various mechanisms are involved including CMV acting as a "teratogen", the effect of immune response and inflammatory mediators, the impact of CMV on apoptosis, and the effect of CMV on the endovascular system [68].

1.3.3 Clinical Features

More than 90% of pregnant women with primary CMV infections are asymptomatic. When symptoms do occur, they are usually non-specific, often described as a "flu-like syndrome" or a "mononucleosis-like illness." Clinical manifestations may include fever, fatigue, headache, myalgia, pharyngitis, and lymphadenitis [69]. Low birth weight and small for gestational age can be seen in infants born to mothers exposed to CMV during pregnancy, even when the infant is not infected [70]. Sonographic fetal anomalies can be detected antenatally in affected fetuses and may include growth restriction, cerebral calcifications, cerebral ventriculomegaly, microcephaly, cerebellar hypoplasia, hyperechogenic fetal bowel, hepatomegaly, ascites, pleural effusion, hydronephrosis, fetal hydrops, and an increase in placental vertical diameter [71, 72]. Magnetic resonance imaging (MRI) has been used in examining fetuses suspected of infection, however, its use is controversial [73].

Ten to 15% of cases are symptomatic and present with the classic CMV inclusion disease. Manifestations of CMV inclusion disease include intrauterine growth restriction (IUGR), hepatosplenomegaly, jaundice, pneumonitis, and purpura. Severe CNS involvement occurs in 50–90% of these infants with microcephaly, intracerebral periventricular calcifications, and progressive sensory neural hearing loss (SNHL). Death may occur secondary to secondary bacterial infection, hepatic dysfunction, bleeding, or disseminated intravascular coagulopathy (DIC). Viral-associated hemophagocytic syndrome or severe end-organ disease of the liver, lungs, bone marrow, or central nervous system can also occur [74]. In survivors, jaundice and hepatosplenomegaly may subside, but neurologic sequelae (e.g., microcephaly, intellectual disability, cerebral palsy, and hearing disorders) persist [75].

Chorioretinitis is the most common ocular abnormality in symptomatic infants and correlates with poor long-term neurodevelopmental outcome. Other findings on eye examination may include retinal (macular) scars, optic atrophy, central vision loss, or strabismus [76]. Cataracts and microphthalmos are not typical in infants with congenital CMV and strongly suggest a disease other than CMV [77].

Endocrinopathies, such as Grave's disease [78] and diabetes insipidus [79], and renal disease, such as nephrotic syndrome [80], have been reported in newborns with symptomatic congenital CMV, but it is unclear whether these conditions are caused by CMV.

Most congenitally CMV-infected infants (85–90%) are subclinical cases. These infants are still at increased risk for SNHL during the first 6 years of life. SNHL is the most common sequelae of congenital CMV and is detected in one-third to one-half of infants with symptomatic disease [56]. Subtle differences, such as lower birth weight and slightly earlier gestational age, have been observed in newborns with asymptomatic congenital CMV infection [81]. Ocular abnormalities, including retinal lesions and strabismus, occur in 1–2% of infants with apparently asymptomatic congenital CMV, but are rarely sight-threatening [76]. Abnormal brain imaging findings of periventricular leukomalacia, ventriculomegaly, and punctate calcifications have been observed in 5–20% of otherwise asymptomatic congenitally-infected newborns [72].

1.3.4   Diagnosis

1.3.4.1 Diagnosis of Maternal Infection

In the setting of maternal infections, prenatal diagnosis can be made through amniocentesis for CMV-DNA PCR in the amniotic fluid, which is most sensitive after 21 weeks of gestation and after 6 weeks from maternal infection/exposure. However, the viral load in the amniotic sac does not appear to be a reliable predictor of neonatal clinical outcome [82]. The only other diagnostic option, cordocentesis, provides similar sensitivity and specificity to amniotic fluid CMV testing, but with a higher complication rate than amniocentesis. Routine screening for CMV infection during pregnancy, whether universal or targeted, is not recommended [61], as it does not meet criteria for an effective screening test at this time, and it can lead to unnecessary intervention which could be harmful, most notably parental anxiety associated with the diagnosis of CMV infection given that approximately 80% of CMV-infected infants will not have neurodevelopmental problems [83].

Recent primary maternal CMV infection may be also diagnosed by the detection of CMV-specific IgM and low-avidity IgG in serum of a previously seronegative mother (seroconversion). The antibody response usually appears at about 2–12 weeks after virus contact. However, it should be noted that the presence of IgM alone should not be used for diagnosis of primary maternal CMV infection as <10% of women with positive IgM congenitally infect their infants, compared with 30–50% of those with seroconversion. This is largely related to the high (up to 90%) false-positive rate for CMV-IgM assays performed by standard enzyme-linked immunoassays (ELISA). Moreover, mothers with reactivation as well as primary infection may have positive CMV-IgM [66, 84]. The use of CMV-specific IgG avidity testing has proved to be helpful in discriminating between primary infection and recurrent infection. The IgG avidity is >80% in women with recurrent CMV infection, however, low IgG avidity is demonstrated for 18–20 weeks after a primary infection [85].

1.3.4.2 Diagnosis of Congenital CMV Infection in the Infant

The gold standard for diagnosis of congenital CMV infection in the infant is through detection of the virus in neonatal urine or saliva cultures obtained before 3 weeks of age. PCR for CMV DNA in blood is as sensitive as a urine culture. Detection of CMV-specific IgM should not be used for diagnosis because it is less sensitive than culture or PCR. Only 70% of CMV-infected neonates have IgM antibodies at birth [86].

The most common ultrasound findings warranting investigation for CMV infection include echogenic fetal bowel, cerebral ventriculomegaly and calcifications, and fetal growth restriction. Hepatic calcifications, microcephaly, and subependymal cysts have also been described [72].

Neuroimaging of the brain using computed tomography (CT) scans, MRI, or cranial ultrasonography demonstrates abnormalities in 70% of infants with symptomatic congenital CMV infection [87] [Table 1.3]. Intracranial calcification is the most frequently reported imaging finding of congenital CMV infection, occurring in 34–70% of patients [88]. CT is highly sensitive for its depiction and localization, although calcification also is depicted at MRI and ultrasonography; ultrasonography may be more sensitive than MRI for the depiction of calcification [89].

Calcification occurs in a variety of locations including periventricular regions (Fig. 1.5) and within the basal ganglia (Fig. 1.6) and brain parenchyma. Although calcification is extremely common in patients with congenital CMV infection, it is not always present; an absence of calcification should not exclude a diagnosis of congenital CMV infection [90].

At CT, the presence of intracranial calcification in patients with congenital CMV infection is associated with developmental delays, and calcification is more strongly

Table 1.3 Ultrasonographic findings in congenital CMV infections

Brain	• Cerebral calcifications
	• Microcephaly
	• Subependymal cysts
	• Cerebral ventriculomegaly
	• Cerebellar hypoplasia
	• Migrational abnormalities e.g. polymicrogyria, pachygyria, and lissencephaly
	• Periventricular leukomalacia
Abdomen	• Echogenic bowel
	• Ascites
	• Hepatomegaly
	• Hepatic calcifications
	• Hyperechogenic kidneys
Heart	• Pericardial effusion

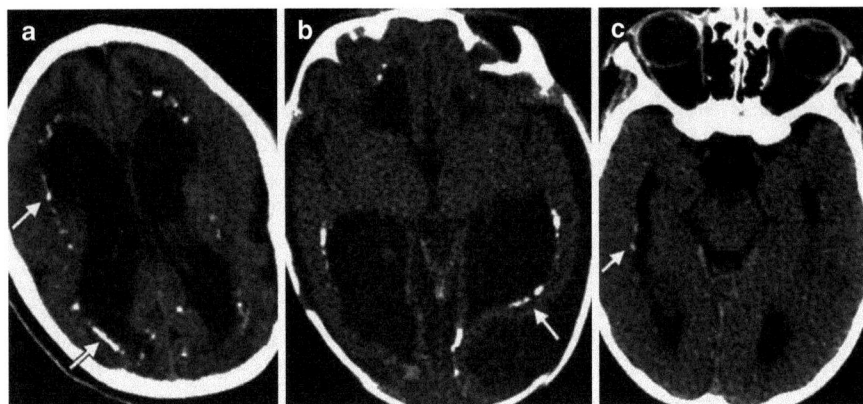

Fig. 1.5 Unenhanced CT images show periventricular calcification (*arrows*) in three patients with early (**a**), late second trimester (**b**), and late perinatal (**c**) CMV infection (Source: [87])

associated with intellectual disability than other abnormal imaging findings [91]. Although calcification is indicative of neurodevelopmental delays, it is not specific, and patients with normal neurologic function may have intracranial calcification related to congenital CMV infection. Moreover, normal brain imaging does not necessarily predict normal neurodevelopmental outcome, particularly since hearing loss is frequently progressive in congenital CMV [92].

Non-specific laboratory findings include thrombocytopenia, hemolytic anemia, and abnormal liver function tests. SNHL may be present at birth or occur later in childhood (congenital CMV is the leading cause of SNHL). The hearing loss associated with symptomatic congenital CMV is often progressive, and eventually becomes severe to profound in the affected ear(s) of 78% of children [77]. Repeated auditory evaluation is recommended.

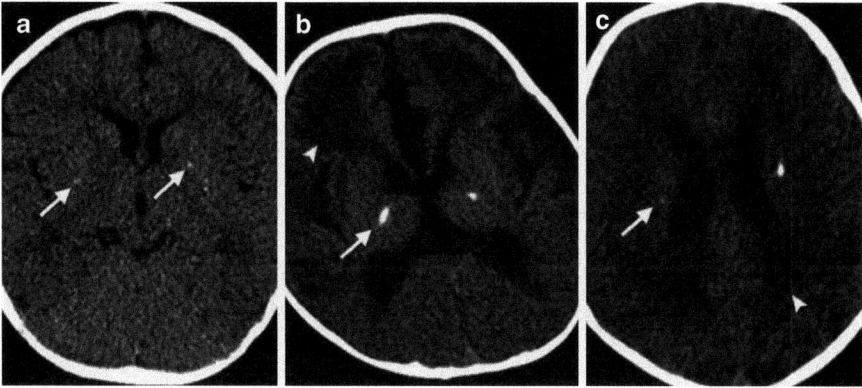

Fig. 1.6 Basal ganglia calcification in three patients with congenital CMV infection. (**a**) Unenhanced axial CT image shows basal ganglia calcification (*arrows*). (**b, c**) Unenhanced axial CT images show calcification (*arrow*), volume loss, ventriculomegaly, and white matter lesions (*arrowhead*), which appear hypoattenuating, findings indicative of congenital CMV infection. In patients with congenital CMV infection, calcification may be subtle, unilateral, and less chunky than that associated with other entities (Source: [87])

1.3.5 Treatment

Treatment should be initiated as soon as virologic testing is confirmed. Severely symptomatic infants should be monitored with quantitative CMV-DNA PCR and prolonging antiviral treatment is to be given if viremia has not resolved after 6 months [62]. No antiviral agent has been yet approved for treatment of congenital CMV infection. CMV differs from other herpes viruses in that it lacks the viral enzyme thymidine kinase (TK), thus it is resistant to antiviral agents that depend on this enzyme for their action, such as acyclovir [93].

Ganciclovir (6 mg/kg per dose every 12 hours for 6 weeks) is indicated for CMV pneumonitis, hepatitis, encephalitis, and chorioretinitis especially in immunosuppressed patients. The value of the previous regimen in the prevention of hearing deterioration in neonates with symptomatic CMV infection has been suggested in a small, randomized trial [94]. Longer duration of therapy (6 months vs. 6 weeks) did not improve hearing in the short term but appeared to improve hearing and developmental outcomes modestly in the longer term outcomes at 12–24 months [95]. Ganciclovir can induce neutropenia (60%), thrombocytopenia, anemia, hyperkalemia, and renal impairment as side effects [96]. These drug-related adverse events are reversible upon discontinuation of the drug [97]. Valganciclovir (a prodrug of ganciclovir) can be an alternative that is administered orally at 16 mg/kg per dose twice daily [95].

Other potentially efficacious therapies for congenital CMV disease are foscarnet and cidofovir which are reserved for refractory CMV infections, ganciclovir resistance or toxicity, and co-infection with adenovirus [98]. Hyperimmune anti-CMV immunoglobulin (CMV-IG) may benefit infants with a fulminant presentation [99].

1.3.6 Prevention

Seronegative pregnant women should protect themselves by following good personal hygiene, especially hand washing with soap and water, and avoiding kissing children under age 6 on the mouth or cheek; instead, they can kiss them on the head or give them a hug. Avoidance of sharing food, drinks, or oral utensils (e.g., spoon, pacifier) with young children is also important. All toys and other surfaces that come into contact with children's urine or saliva should be cleaned. CMV seronegative blood and blood products should be used if a blood transfusion is indicated [100, 101].

Passive immunization of expectant mothers with hyperimmune anti-CMV immunoglobulin (CMV-IG) following primary CMV infection (prophylactic use) was shown to lower the risk of maternal-fetal transmission of CMV [102]. Until further scientific evidence is available from the results of ongoing, multicenter, prospective, randomized clinical trials become available, the use of CMV-IG in pregnant women with primary CMV infection is not generally recommended, and its use should be individualized [83]. A number of CMV vaccine candidates are under investigation [103].

Administration of valacyclovir to pregnant women with confirmed fetal CMV infections decreased fetal CMV viral loads and provided therapeutic concentrations of drug in the maternal and fetal circulations. Approximately 50% of fetuses whose mothers received treatment were developing normally at ages 1–5 years [104]. However, antenatal treatment with ganciclovir or valacyclovir is not recommended by many experts and should only be offered as part of a research protocol [83, 105].

1.3.7 Prognosis

Congenital CMV infection is the most common non-genetic cause of childhood SNHL, detected in one-third to one-half of infants with symptomatic disease [64]. The CMV-infected infant is at increased risk for many other neurodevelopmental disabilities such as visual impairment, and intellectual disability. The presence of clinical symptoms at birth and high CMV viral load are the best predictors of developmental disabilities [106]. In symptomatic infants with congenital CMV infection, mortality rates range from 4 to 30%, and at least 90% of them have late sequelae [107].

Moreover, about 7–15% of asymptomatic congenitally infected infants will develop SNHL [108]. Early identification of a sensory or developmental disability with subsequent intensive interventions has been shown to improve clinical outcomes [106]. Only one third of infants with congenital CMV infection who eventually develop hearing loss are identifiable with newborn hearing screening at birth; hearing loss is often progressive, and eventually becomes severe to profound in the affected ear(s), thus, hearing evaluations are recommended every 3–6 months during the first 3 years of life, and annually thereafter until at least 18 years of age. If hearing loss is detected, assessments should continue into adulthood to monitor for hearing loss progression [109].

Strabismus and amblyopia may develop, therefore, these infants should have close ophthalmologic follow-up throughout their lives [76]. Poor long-term neuro-developmental outcome is associated with microcephaly, chorioretinitis, and abnormal head CT scan findings, particularly intracranial calcifications [91]. Treatment with antiviral medications in symptomatic infants in early infancy appears to reduce long-term sequelae, particularly hearing loss [95]. Infants with asymptomatic infection at birth are less likely to develop long-term disabilities [110].

1.4   Perinatal Herpes Simplex Virus Infection

The two types of the herpes simplex virus (HSV), HSV-1 and HSV-2, are members of the Herpesviridae family of DNA viruses, grouped along with varicella-zoster virus in the alphaherpesvirus subfamily. The alphaherpesviruses are characterized by short reproductive cycles, host cell destruction during active replication, and the ability to establish lifelong latency in sensory neural ganglia [111]. Both HSV-1 and HSV-2 are large virions with a lipid envelope surrounding an icosahedral nucleocapsid [112].

HSV-1 and HSV-2 infections are common in both developed and less-developed countries and are associated with life-long infection. Although either HSV-1 or HSV-2 can cause genital infection, HSV-1 has typically been more associated with orolabial lesions whereas HSV-2 has historically been the more common cause of genital lesions [112].

Neonatal HSV infection is uncommon, affecting 1 out of every 3200 to 10,000 live births, however, it is associated with high mortality and morbidity [113, 114]. The incidence of neonatal HSV infection has increased dramatically over the past several decades [115, 116]. It is critical to recognize that most mothers of infants with neonatal HSV do not have a history of HSV infection themselves (due to asymptomatic shedding or atypical lesions).

1.4.1   Routes of Transmission

Maternal infection with HSV occur through inoculation of oral, genital, or conjunctival mucosa or breaks in skin with the virus, which then infects the sensory nerve endings, and which in turn is conveyed by retrograde axonal flow to the dorsal root ganglia, where the virus develops lifelong latency. Latent HSV is not susceptible to antiviral drugs, and maternal infection, even after antiviral therapy, is lifelong [113]. Several risk factors for HSV genital infection have been identified, including female gender, longer duration of sexual activity, minority ethnic group, prior history of other genital infections, lower family income, and number of sex partners [117].

Mother-to-child transmission of HSV can occur in-utero (5%), during the peripartum period (85%), or postnatally (10%) [118]. HSV is not transmitted through breast milk; rather, postnatal acquisition of HSV is due to direct contact with a person shedding the virus, usually via an orolabial or other cutaneous lesion. Possible routes for postnatal transmission include oropharyngeal shedding by either parents, hospital personnel, or other contacts as well as active maternal breast lesions.

Maternal antibody to HSV is associated with decreased risk of fetal and neonatal transmission. In a US study population, an increased transmission risk of 57% in women with first-episode primary genital infections is demonstrated versus a 2% risk increase in those with recurrent genital lesions [119]. Other factors associated with increased risk of mother-to-child HSV transmission include detection of HSV-1 or HSV-2 from the cervix or external genitalia via viral culture or PCR, duration of rupture of membranes, disruption of the neonate's cutaneous barrier by the use of a fetal scalp electrode or other invasive instrumentation, and vaginal delivery [116].

1.4.2   Clinical Features

Antenatal transmission of HSV (in-utero infection) is uncommon but may lead to spontaneous abortion, hydrops fetalis, and in-utero fetal demise. Congenital intrauterine infection, that usually is identified within the first 48 h following birth, is characterized by skin vesicles, scarring, hypo- or hyperpigmentation, eye lesions (chorioretinitis, microphthalmia, cataract, keratoconjunctivitis, or optic atrophy), neurologic damage (intracranial calcifications, microcephaly, seizures, encephalomacia, or hydranencephaly), growth and psychomotor development delay [120].

Infants infected intrapartum or postnatally by HSV can be divided into three major categories: (1) HSV disease localized to the skin, eye, and/or mouth (SEM disease); this syndrome is associated with a low mortality but it has a significant morbidity and it may progress to encephalitis or disseminated disease if left untreated [121]; (2) HSV encephalitis with or without skin, eye, and/or mouth involvement which causes neurologic morbidity among the majority of survivors [122, 123]; (3) disseminated HSV which manifests as severe multi-organ dysfunction (including central nervous system, liver, lung, brain, adrenals, skin, eye and/or mouth) and has a mortality risk that exceeds 80% in absence of therapy [124]. The various presentations of HSV infection in neonates are summarized in Table 1.4.

Table 1.4 Neonatal herpes simplex virus infection disease classification

Skin, eye, mouth (SEM) disease	• Lesions on the skin, eye, or mouth
	• Without evidence of systemic and/or CNS infection
Central nervous system disease (HSV meningoencephalitis)	• Hypotonia
	• Seizures
	• Abnormal brain imaging
	• Abnormal electroencephalogram (EEG)
	• Cerebrospinal fluid (CSF) pleocytosis and/or proteinosis
Disseminated disease	• Hepatitis
	• Pneumonitis
	• Disseminated intravascular coagulopathy (DIC)
	• With or without evidence of CNS disease

1.4.2.1 Skin, Eye, Mouth (SEM) Disease

SEM disease is the most common and represents 45% of neonatal HSV infections. It usually occurs in the first 2 weeks of life. A cluster of vesicles with an erythematous base are the hallmark of this type of infection (>80%) which, usually appear at the presenting part of the body, or at sites of localized trauma, such as scalp monitor sites. Ocular affection may present early with excessive watering of the eye, crying, and conjunctival erythema with or without periorbital skin vesicles. It may progress to keratoconjunctivitis, chorioretinitis, cataracts, and retinal detachment. Oropharyngeal manifestations include localized ulcerative lesions of the mouth, palate, and tongue. SEM disease usually has a favorable prognosis if treated early. However, in the absence of treatment, it may progress to a central nervous system or disseminated disease [116, 125].

1.4.2.2 Central Nervous System (CNS) Disease (HSV Meningoencephalitis)

CNS disease accounts for approximately 30% of neonatal HSV. It usually occurs between the second and third week of life and is probably the result of a retrograde spread from the nasopharynx and olfactory nerves to the brain. Clinical manifestations include seizures (focal and generalized), fever, lethargy, irritability, tremors, poor feeding, temperature instability, and bulging fontanelle. Skin vesicles may appear later, particularly in untreated infants. Cutaneous lesions can be a diagnostic clue as about 60–70% of neonates with HSV CNS disease have skin vesicles during the course of the disease. However, as many as 35% of infants with HSV CNS disease will never have a vesicular rash identified; thus, as with disseminated HSV disease, the absence of a rash does not rule out neonatal HSV infection [122, 123]. CNS disease may occur with or without SEM involvement and with or without disseminated disease, making it indistinguishable from other causes of neonatal sepsis or meningitis. Death rarely occurs if treated early, however, severe neurologic impairment is frequent [113, 126].

1.4.2.3 Disseminated HSV Disease

Disseminated HSV disease accounts for 25% of neonatal HSV disease. These neonates usually present in the first week of life, with subtle, non-specific sepsis-like clinical manifestations, such as hypothermia or hyperthermia, irritability, lethargy, poor feeding, apnea, respiratory distress, abdominal distension, hepatomegaly, and ascites. Nearly every organ may be involved; the liver (hepatitis, direct hyperbilirubinemia and ascites, may progress to liver failure), the lungs (pneumonia and hemorrhagic pneumonitis, with or without effusion, may progress to respiratory failure), CNS (meningoencephalitis), heart (myocarditis, myocardial dysfunction, hypotension and shock), hematological (DIC, thrombocytopenia, and neutropenia), adrenal glands, kidneys, gastrointestinal tract (necrotizing enterocolitis), skin and mucous membranes lesions, which may appear late in the course of disseminated HSV disease; however, 20–40% of these neonates do not have skin lesions.

Disseminated HSV is often indistinguishable at its onset from both neonatal enterovirus infection and bacterial sepsis. Thus, laboratory evaluation for HSV and empiric antiviral therapy should be considered in high-risk neonates with a sepsis-like picture, meningoencephalitis, progressive pneumonitis, or hepatitis. Disseminated infection has the worst prognosis, with a mortality of about 80–90% if not treated and neurodevelopmental impairment in about 20% of survivors [122, 127].

1.4.3 Diagnosis

Isolation of HSV by culture remains the definitive diagnostic method of establishing neonatal HSV disease [128]. If skin lesions are present, a scraping of the vesicles should be transferred, in appropriate viral transport media on ice, to a diagnostic virology laboratory. Other sites from which specimens should be obtained for culture of HSV include the conjunctivae, mouth, nasopharynx, and rectum ("surface cultures"). Culture results are usually established in 1–3 days. HSV isolated in cell culture can be typed as HSV-1 or HSV-2 using type-specific immunofluorescence assays. Culture enhancement systems, such as shell vial centrifugation can provide rapid detection of HSV in clinical systems within 24–48 hours [129].

Rapid diagnostic techniques are available, such as direct fluorescent antibody staining of vesicle scrapings or enzyme immunoassay detection of HSV antigens. These techniques are as specific, but slightly less sensitive, than culture [128].

The diagnosis of neonatal HSV CNS disease has been greatly enhanced by PCR testing of cerebrospinal fluid (CSF) specimens and PCR assay is now the method of choice for documenting CNS involvement in an infant suspected of having HSV disease. CSF PCR assay for HSV DNA should be conducted in all neonates with suspected or proven neonatal HSV infection, whatever its type, as approximately 25% of neonates with SEM disease, and more than 90% of neonates with disseminated HSV disease, have positive HSV DNA results in their CSF samples. Similarly, detection of HSV DNA in the blood or plasma confirms the diagnosis in neonates with disseminated HSV disease [130, 131]. All infants with a positive CSF PCR assay result for HSV DNA at the beginning of antiviral therapy should have a repeat lumbar puncture near the end of treatment to determine that HSV DNA has been cleared from the CNS. Infants whose PCR assay result remains positive should continue to receive intravenous antiviral therapy until the CSF PCR assay result is negative [132].

With herpes encephalitis, high CSF protein level and pleocytosis may be seen. The EEG may show temporal lobe changes and the brain CT/MRI can also detect patchy calcification in the temporal area. In patients with disseminated disease, additional laboratory tests should be performed to assess the degree of organ involvement and exclude other conditions with similar manifestations. These tests include complete blood count, total and direct bilirubin, liver enzymes, ammonia, blood urea nitrogen, creatinine, urine analysis, CSF cytochemistry, blood and CSF cultures. A diffuse interstitial pattern is observed on radiographs of infants with HSV pneumonitis. In neonates with HSV hepatitis and acute liver failure, ultrasonography may reveal hepatomegaly and ascites [129].

1.4.4 Treatment

Acyclovir is the antiviral agent of choice for the treatment of all types of neonatal HSV infections. [133]. Acyclovir therapy is indicated in virologically-proven HSV infections, as empirical therapy in clinically suspected cases pending viral studies, and as a preventive measure in asymptomatic neonates at risk due to exposure (maternal active HSV genital lesions) [129].

The full course of acyclovir therapy should be administered for infants with positive HSV cultures or PCR, and infants with negative virologic studies in whom neonatal HSV is strongly suspected. Infants with SEM disease should receive acyclovir 20 mg/kg per dose every 8 hours intravenously for 14 days, and those with CNS or disseminated disease should receive the same dose for 21 days or longer if the CSF PCR remains positive. Administration through central access or a peripherally-inserted central catheter is preferred. Acyclovir therapy should be continued while clinical observation and results of laboratory and imaging evaluations are completed. This dose should be adjusted in neonates with impaired kidney functions. Infants with ocular involvement are treated with topical ophthalmic agents (1% trifluridine, 0.1% iododeoxyuridine, or 3% vidarabine) in addition to parenteral therapy.

Neurologic outcomes for infants with CNS disease have recently been improved upon with use of oral suppressive acyclovir therapy administered for 6 months following the completion of a standard acyclovir neonatal HSV treatment course. Infants with CNS disease who received suppressive acyclovir therapy at a dose of 300 mg/m^2/dose administered orally three times a day for 6 months had better neurodevelopmental outcomes compared to the placebo group, and infants with CNS and SEM disease had less frequent recurrences of skin lesions while receiving the suppressive therapy [133].

Supportive measures for the critically ill neonate with disseminated or CNS disease include fluid and electrolyte maintenance, avoidance of hypoglycemia, management of shock and systemic inflammatory response, provision of oxygen and mechanical ventilator support, control of seizures, nutritional support, management of disseminated intravascular coagulation, including fresh frozen plasma and/or platelet transfusions, antimicrobial treatment for secondary bacterial infections, and IV immune globulin (IVIG), especially if myocardial dysfunction/ myocarditis or systemic inflammatory response is present. Glucocorticoids, exchange transfusion, and liver transplantation have been shown to effective in case reports [134, 135].

1.4.5 Management of Asymptomatic Neonates Who Are Exposed to HSV During Delivery

The management of asymptomatic neonates who are exposed to HSV during delivery depends upon the mother's history of genital HSV, maternal serology for type specific HSV, and the presence or absence of HSV genital lesions at the time of delivery. All such infants should be monitored for evidence of HSV during the first 6 weeks of life. Neonates who develop clinical evidence of HSV infection should undergo full virologic evaluation and be started on IV acyclovir, while results of virologic evaluation are pending.

In asymptomatic neonates born after vaginal or cesarean delivery to women with lesions at delivery and history of genital HSV preceding pregnancy. In these conditions, the possibility that these active lesions represent reactivation of latent HSV is high, and, therefore, the likelihood of transmission to the infant is low (2%). Skin and mucosal specimens should be obtained from the neonate for culture at approximately 24 hours after delivery, and blood should be sent for HSV-DNA PCR assay. If the infant remains asymptomatic, there is no need for acyclovir therapy and the infant should be observed in the hospital until HSV cultures are negative. If the surface and blood virologic study results are negative at 5 days, the infant should be evaluated if signs or symptoms of neonatal HSV disease develop during the first 6 weeks of life. On the other hand, if the surface and blood virological study results become positive, thus confirming neonatal HSV infection, the infant should undergo a complete evaluation (lumbar puncture with CSF sent for indices and HSV-DNA PCR assay, in addition to serum alanine transaminase) to determine the extent of disease, and IV acyclovir should be initiated as soon as possible. If the evaluation findings are normal, indicating that the neonate has HSV infection but that it has not yet progressed to HSV disease, the infant should be treated empirically with IV acyclovir for 10 days to prevent progression from infection to disease. If the evaluation findings are abnormal, indicating the neonate already has neonatal HSV disease, the infant should be treated accordingly [128].

For asymptomatic neonates born after vaginal or cesarean delivery to women with lesions at delivery and no history of genital HSV preceding pregnancy, the presence of genital lesions at delivery could represent first-episode primary infection (risk of transmission to the newborn infant of 57%), first-episode non-primary infection (risk of transmission to the newborn infant of 25%), or recurrent infection (risk of transmission to the newborn infant of 2%). Serological testing should be performed to determine the type of HSV infection in the mother. At approximately 24 hours after delivery, neonatal mucocutaneous specimens should be obtained for culture, blood for HSV-DNA PCR, CSF cytochemistry, HSV-DNA PCR and serum alanine transaminase and IV acyclovir should be initiated pending the results of the evaluation. If the mother has a documented or assumed first-episode primary or first-episode non-primary infection and the neonate's evaluation results are normal, the infant should be treated empirically with IV acyclovir for 10 days to prevent progression from neonatal infection to disease (preemptive therapy). If the infant's evaluation results are abnormal, indicating the neonate already has neonatal HSV disease, the infant should be treated accordingly. If the mother has recurrent genital HSV infection (that is, she has type-specific antibody to the HSV serotype detected by her genital swab) and the neonatal surface and blood virology study results are negative, acyclovir should be stopped, and the infant may be discharged and reevaluated with any signs or symptoms of illness during the first 6 weeks of life. Conversely, if virology study results are positive, confirming neonatal HSV infection, the infant should receive treatment accordingly. IV acyclovir should be continued, with the duration of treatment based on the evaluation for neonatal HSV disease. If the evaluation findings are normal, the infant should be treated empirically with IV acyclovir for 10 days to prevent progression from HSV infection to HSV disease. If the evaluation findings are abnormal, the infant should be treated accordingly [128].

1.4.6 Prevention

Since the majority of mother-to-child transmission of HSV infection occurs as a result of exposure to virus shed from the genital tract as a neonate passes through the birth canal, the most common strategies for preventing transmission seek to reduce neonatal exposure to active genital lesions [112]. In women with active recurrent genital HSV lesions, antiviral suppressive therapy with oral acyclovir or valacyclovir can be started at 36 weeks of gestation, a practice that has been associated with decreased genital lesions at the time of delivery, decreased viral detection by viral culture or PCR, and subsequently a reduced need for cesarean delivery for the indication of genital HSV [136]. Suppressive therapy does not completely obviate the risk of perinatal transmission [137].

Cesarean delivery should be performed in women with active genital lesions (whether primary or recurrent) or with prodromal symptoms that may indicate an impending genital outbreak [117]. For maximum effect on risk reduction, cesarean delivery should be performed prior to rupture of membranes. If rupture has already occurred and genital lesions are present, cesarean delivery is still recommended. Cesarean delivery is not indicated in women with a previous diagnosis of genital herpes, and had no genital lesions at the time of labor [138].

Infants and mothers with HSV lesions should be in contact isolation applying careful hand washing. Contact precautions should be applied also to asymptomatic infants born to women with active herpetic genital lesions until the end of the incubation period (day 14) or until samples from the infant's oropharynx/nasopharynx and conjunctivae taken after the first 24 hours of life are negative; this practice is not necessary if a cesarean section (with membranes rupture <4 hours) is performed [129]. Acyclovir need not be started as long as the infant remains asymptomatic [133].

Women with herpetic breast lesions should not breastfeed directly from the affected breast until the lesions have resolved, and women with oral HSV should wear a mask while breastfeeding. Nursery personnel with herpetic whitlows should not care for newborns even with gloves [129]. The American College of Obstetricians and Gynecologists (ACOG) does not recommend routine screening of asymptomatic women for HSV during pregnancy [138].

There is no licensed, effective vaccine against HSV-1 or HSV-2 infection. However, a candidate HSV-2 gD subunit vaccine has been tested in a randomized controlled trial, and was found to have an efficacy of 58% for preventing HSV-1 but not HSV-2 genital herpes [139].

1.4.7 Prognosis

The outcome of neonatal HSV infection depends largely upon the clinical presentation and promptness of treatment. Disseminated disease has the worst prognosis with a 1-year mortality rate of 29% [123]. Risk factors associated with mortality include prematurity and the presence of severe hepatitis, acute liver failure, pneumonitis, lethargy, coma, and DIC [122, 140]. About 20% of these infants develop neurodevelopmental abnormalities e.g., developmental delay, hemiparesis, persistent seizures, microcephaly, blindness [118]. Intravenous acyclovir therapy has improved mortality rates dramatically, but had less effect on morbidity [122].

The 1-year mortality rate for CNS disease is 4%. Risk factors of mortality in these infants include prematurity, seizures, and coma at the time of presentation. About 70% of survivors of neonatal CNS HSV have neurodevelopmental abnormalities [118, 123, 140].

Infants with SEM HSV disease typically have minimal mortality and lower morbidity. Less than 2% of neonates treated with acyclovir have neurodevelopmental abnormalities [118, 123, 140]. The risk of neurodevelopmental abnormalities is increased in infants with ≥3 recurrences of skin lesions before 6 months of age [140]. Patients with ocular involvement are at risk for long-term complications, including vision loss, and require close follow-up. Early antiviral treatment of infants with SEM disease prevents progression to CNS or disseminated disease [141].

The use of intravenous acyclovir at a dose of 60 mg/kg/day in three divided doses has improved 1-year mortality rates to 29% and 4% for disseminated and CNS diseases, respectively [123]. Furthermore, this 60 mg/kg/day dose of acyclovir was shown to improve neurodevelopmental outcomes in infants with disseminated disease (83% of survivors had normal neurodevelopmental outcomes), but not for infants with CNS disease (31% of survivors had normal neurodevelopmental outcomes). Further studies aimed at optimizing the efficacy of antiviral therapy, including the development of novel antiviral agents and the investigation of the potential benefit of combination therapies, are needed [112].

Even after successful IV acyclovir therapy, recurrence of HSV can occur and may be a lifelong problem for the patient and family. Recurrent mucocutaneous lesions are common and occur in 50–80% of neonates, with 1–12 episodes in the first year of life [129]. However, infants with ≥3 cutaneous recurrences during the first 6 months of life are at increased risk of neurodevelopmental abnormalities [140]. On the other hand, recurrence of CNS disease is rare [142]. Long-term suppressive therapy with oral acyclovir is recommended to reduce skin or eye recurrences during infancy.

1.5 Parvovirus B19 Infection

Human parvovirus B19 (PB19) is a small, single-stranded, non-enveloped DNA virus. The only known natural host cell is the human erythroid progenitor cell. PB19 is a potent inhibitor of hematopoiesis [143]. Pregnant women lacking antibodies to the virus are as susceptible as any other immunocompetent adult to PB19 infection. However, 35–53% of pregnant women have preexisting IgG to the virus, indicating immunity from a prior infection. The incidence of acute PB19 infection in pregnancy is 3.3–3.8% [144]. Infection with PB19 usually confers lifelong immunity [145].

1.5.1 Routes of Transmission

Maternal infection with PB19 virus is most commonly spread by respiratory secretions or from hand to mouth contact, but the virus can also be transmitted by blood or blood products [146]. Women at increased risk include mothers of preschool and school age children, workers at daycare centers, and school teachers [147]. The transmission rate of maternal PB19 infection to the fetus is 17% to 33% [148]. B19 also may stimulate a cellular process initiating apoptosis (programmed cell death) [149]. The latter may account for the minimal inflammatory response noted in tissues infected with B19.

1.5.2 Clinical Features

Maternal infection during pregnancy may be asymptomatic or present with nonspecific flu-like syndrome of low-grade fever, sore throat, generalized malaise, and headache. Pregnant women rarely develop the characteristic "slapped cheek" appearance that is seen among school-aged children where the major manifestation is erythema infectiosum (fifth disease) [150].

Arthropathy affects up to 50% of pregnant women with PB19 infection and may last several weeks to months [148]. This immune mediated arthopathy presents as symmetric polyarthralgia, affecting the hands, wrists, ankles, and knees [147].

Maternal PB19 infection in the first trimester most commonly results in fetal loss or miscarriage [151]. Later, fetal infection leads to marked fetal anemia, high-output cardiac failure, and non-immune hydrops fetalis [152]. The risk of developing anemia and fetal hydrops appears to be greater in women infected during the first half of pregnancy [153]. Hydrops can lead to rapid fetal death (within a few days to weeks) or can resolve spontaneously with an apparently normal neonate at birth [154]. Fetal myocarditis, and less often hepatic infection, may contribute to fetal cardiac failure [155]. Severe thrombocytopenia is reported in 37% of PB19-infected fetuses with hydrops [156]. This must be taken into account when the decision is made to perform a cordocentesis or intrauterine transfusion [147]. PB19 itself, in the absence of hydrops or significant fetal anemia, does not seem to cause long-term neurological sequelae, but severe anemia and fetal hydrops may be an independent risk factor for long-term neurological morbidity [157].

Isolated fetal pleural or pericardial effusions that resolve spontaneously before term have been also reported with maternal PB19 infection. These transient effusions may be the result of direct pleural or myocardial inflammation [158].

Despite case reports suggesting a link between PB19 infection during pregnancy and fetal malformations [159, 160], epidemiologic studies do not support this association [161]. PB19 is probably not a teratogen [162].

1.5.3 Diagnosis

Exposed pregnant woman should have their PB19 IgG and IgM status determined. Blood PCR for PB19 DNA is indicated when the fetus is hydropic. In contrast to maternal testing, serologic examination (PB19 IgG and IgM antibodies) of fetal and neonatal blood samples is unreliable because the fetus does not begin to make its own IgM until 22 weeks' gestation [163]. On the other hand, PCR to detect PB19 DNA is extremely sensitive in neonates [164]. PB19 cannot usually be cultured in regular culture media. The virus can be identified histologically by characteristic intranuclear inclusions or by the presence of viral particles on electron microscopy [165].

Antenatal ultrasound is used to monitor for hydrops development and fluid accumulation in fetal body cavities. Doppler velocimetry is used to detect increased fetal middle cerebral artery peak systolic velocity (MCA-PSV); a very sensitive measure of fetal anemia [166].

Elevated maternal serum alpha-fetoprotein (MSAFP) levels have been associated with fetal PB19 infection in several case reports [167]. However, it cannot be used as a reliable marker of fetal infection [145].

1.5.4 Management

Pregnant women who are exposed to, or have symptoms of parvovirus infection, and are found to be immune (IgG positive, IgM negative) can be reassured that recent exposure will not result in adverse consequences. If there is no immunity to the virus and seroconversion has not occurred after 2 weeks, the woman is not infected but remains at risk. If the woman has been infected (IgM positive), the fetus should be monitored for development of hydrops fetalis by ultrasound and Doppler assessment of MCA-PSV, weekly until 8–10 weeks postexposure [147].

Intrauterine blood transfusion (IUT) is indicated for hydrops and/or symptomatic fetal anemia [168]. However, this procedure is not feasible before about 20 weeks of gestation due to limited visualization and the small size of the relevant anatomic structures [169]. When preparing for fetal transfusion, both packed red blood cells and platelets must be available because some fetuses have severe thrombocytopenia in addition to anemia, as discussed earlier.

If delivery of a hydropic or anemic infant is planned, this should occur in a tertiary care center with staff and resources to manage these neonates. The use of corticosteroids to accelerate lung maturity is not contraindicated [147]. Resuscitation of such infants is frequently difficult and advance preparation is advisable. The majority of hydropic infants require respiratory assistance and mechanical ventilation. Ventilation may be compromised by pulmonary hypoplasia, pulmonary edema, air leaks, or by the accumulation of pleural or peritoneal fluid. Abdominal paracentesis and thoracocentesis of fetal ascites and pleural effusions may be needed, either just prior to delivery, or immediately after to facilitate resuscitation [170].

Postnatal management depends upon the gestational age of the infant, other associated conditions (e.g., respiratory distress syndrome), and illness severity. Infants with severe anemia and cardiovascular instability may benefit from an isovolumetric or partial exchange transfusion with packed red blood cells [170]. No antiviral agents are effective against PB19 virus [171].

1.5.5 Prevention

The best way, currently available, to prevent PB19 infection is to interrupt transmission by adequate infection control practices, such as hand hygiene [172]. Patients with normal immune systems are probably not infectious after the onset of PB19-associated rash, arthralgias, or arthritis; all represent autoimmune manifestations of the infection. During outbreaks, the CDC recommends that pregnant women be informed about the risks of infection and make their own decision as to avoiding the workplace or outbreak settings, after consultation with healthcare workers, public health officials, and family members [173]. Post-exposure prophylaxis with IVIG for pregnant women or patients with chronic hematologic disorders is not indicated. Clinical trials of vaccines against PB19 have been suspended early due to potential vaccine-associated adverse events [174].

1.6 Congenital Varicella Syndrome

Varicella-zoster virus (VZV), also known as human herpes virus 3, is a DNA virus that causes chickenpox (varicella). After primary infection, VZV can remain dormant in the sensory nerve ganglia. Herpes zoster (shingles) can result from the reactivation of the virus in the elderly or immunocompromised individuals, and is characterized by painful vesicular skin eruptions in the sensory nerve roots [175]. VZV has teratogenic potential and is associated with low birth weight [176]. It can also cause a rare but serious multi-system fetal anomaly called the congenital varicella syndrome (CVS).

1.6.1 Route of Transmission

Chickenpox is highly contagious. Maternal infection occurs via respiratory droplets or direct contact with lesions. Herpes zoster (shingles) is much less contagious and usually requires close contact or exposure to open cutaneous lesions for transmission to occur [177]. The incubation period for VZV ranges from 10 to 21 days [175].

Viremia results, before the onset of rash, with transplacental passage to the fetus. Most transplacental infections are asymptomatic but they can cause CVS or neonatal chickenpox, which have high rates of mortality and morbidity.

Most cases of congenital varicella syndrome occur in infants whose mothers were infected between 8 and 20 weeks gestation. However, the overall risk of infection is quite small compared with numerous other viruses acquired during pregnancy. The risk appears to be approximately 2% if the infection occurs before 20 weeks and less than 1% if it occurs before 13 weeks [178].

1.6.2 Pathogenesis

Primary infection with VZV begins with mucosal inoculation of the virus through respiratory exposure or direct contact with fluid from varicella or herpes zoster skin lesions. The incubation period is between 10 and 21 days after contact. VZV had been presumed to replicate first in the regional lymphoid tissues, followed by primary viremia and viral replication in the liver and spleen; a secondary viremia then occurs just before the appearance of skin or mucosal lesions [179]. When primary VZV infection occurs during pregnancy, the virus can be transferred across the placenta to the developing fetus. Infection of the genital tract mucosa leading to an ascending infection, with transfer across the amniotic membranes, may also occur [180]. Nosocomial acquisition of VZV in the neonatal intensive care unit (NICU) also can occur specially in premature infants, as active transfer of maternal IgG antibodies occurs primarily during the third trimester of pregnancy [181].

One hypothesis is that CVS is caused by reactivation of VZV in utero rather than by primary infection of VZV [182]. Immature fetal cell-mediated immunity would shorten the latent period between primary infection and reactivation [183]. The dermatomal distribution of the skin lesions [176], segmental maldevelopment and dysfunction of the affected system support this hypothesis [182].

1.6.3 Clinical Features

Characteristic findings of affected infants include some or all of the following, in order of the frequency of occurrence, cicatricial skin lesions, which may be depressed and pigmented in a dermatomal distribution; ocular defects, such as cataracts, chorioretinitis, Horner syndrome (ptosis, miosis, and enophthalmos), microphthalmos, and nystagmus; limb abnormalities that affect almost half of the cases and often include hypoplasia of bone and muscle; central nervous system abnormalities, such as cortical atrophy, seizures, and intellectual disability. Intrauterine growth restriction commonly occurs [184]. Presentation of zoster during early infancy is common in affected children. The clinical manifestations of CVS are summarized in Table 1.5.

Table 1.5 Clinical manifestations of congenital varicella syndrome (CVS)

Skin	Cicatricial lesions
	Cutaneous defects
	Hypopigmentation
Neurological	Cortical atrophy/porencephaly
	Seizures
	Intellectual disability
	Intrauterine encephalitis
	Autonomic instability
Eye	Chorioretinitis
	Cataracts
	Microphthalmos
	Horner syndrome
	Nystagmus
Musculoskeletal	Hypoplasia of bone and muscle
Gastrointestinal	Duodenal stenosis
	Dilated jejunum
	Gastroesophageal reflux
	Small left colon
	Intestinal atresia
	Hepatic calcifications
Genitourinary	Hydroureter
	Hydronephrosis
	Neurogenic bladder
Others	Intrauterine growth retardation
	Developmental delay
	Cardiovascular defects

1.6.4 Diagnosis

1.6.4.1 Prenatal Diagnosis

Both the PCR of the amniotic fluid or fetal blood for detecting VZV DNA and the serology tests that detect VZV-specific antibodies (varicella-specific IgM or persistence of varicella-specific IgG beyond 7 months of age) are useful in confirming fetal infection, but neither is useful for detecting CVS [185, 186]. Therefore, a detailed ultrasonography, to detect limb deformity, microcephaly, hydrocephalus, polyhydramnios, soft tissue calcification and intrauterine growth restriction [187], is combined with the PCR and serology tests.

1.6.4.2 Congenital Varicella Syndrome in the Neonate

The diagnosis of varicella usually relies on clinical examination. VZV may be cultured from vesicular fluid, but the VZV takes several weeks to grow. Other rapid diagnostic studies include DFA and PCR. PCR is a highly sensitive and specific test that detects VZV from either vesicular swabs or scrapings, scabs from crusted lesions, tissue from biopsy samples, and CSF [175, 188]. Scrapings of skin lesions, as with HSV infections, may show large multinucleated cells when stained with Wright or Giemsa stain (Tzanck smears), but this procedure is not commonly performed.

Serologic tests are of little or no value for the diagnosis of varicella. VZV IgG antibodies only become detectable as early as 7 days after varicella symptoms appear, VZV IgG ELISA methods have substantial false negative rates (15–20%). IgM antibody may be detected as soon as 3 days after VZV symptoms appear, however, no validated serologic tests for VZV IgM are available, only about 25% of infants reported with classic CVS have positive VZV IgM and VZV IgM antibodies can be detected in individuals who have had varicella at some time in the past [189, 190]. Other diagnostic tests, including fluorescent anti-membrane antibody (FAMA), latex agglutination (LA), enzyme-linked immunosorbent assay (ELISA), and complement-enhanced neutralization, are available. All of these tests are more sensitive than is the older complement fixation (CF) assay. Although PCR is the most sensitive, LA and DFA provide the most rapid results. Neuroradiographic demonstration of intracranial calcifications has been reported but is not common [191].

1.6.5 Management

Oral acyclovir therapy (800 mg five times per day for 7 days) is recommended for all pregnant women with uncomplicated varicella. Acyclovir treatment is most effective within the first 24 hours and appears to be safe [192].

Varicella pneumonia during pregnancy is a medical emergency; the mortality rate among pregnant women in the era prior to antiviral therapy approximated 36–40% in case series reports. IV acyclovir (10 mg/kg every 8 hours) can be tried for pregnant women with varicella pneumonia [193]. The risk-benefit of treatment of maternal varicella infection outweighs any theoretical concerns regarding fetal toxicity; no specific pattern of congenital malformations has been attributed to acyclovir in large pregnancy registries [194].

Counseling for elective abortion is problematic because the incidence of the syndrome is so low as to preclude routinely recommending abortion. Maternal infection may have a 25% risk of fetal varicella, but this does not mean that the fetus will develop CVS. Likewise, as discussed earlier, chorionic villus sampling with PCR to detect viral DNA and cordocentesis to show fetal varicella-specific IgM indicate only fetal infection, not the development of the congenital syndrome. Unfortunately, abnormalities on prenatal fetal ultrasound may be detected late enough to preclude an elective abortion. Limb defects carry a 50% risk of intellectual disability and early death [195].

Neonatal management includes the use of acyclovir to stop progression of eye disease or to treat recurrent zoster (shingles), which is common in the first 2 years of life [195]. Associated anomalies are treated as indicated in every case.

1.6.6 Prevention

The VZV vaccine is a live attenuated virus that is not secreted in breast milk. Immunity to chickenpox usually results from natural infection. Therefore, vaccination is needed in only 3.9% of adults [196]. Varicella immunization is recommended for all non-immune women. Post-exposure vaccination during pregnancy or vaccination 3 months before pregnancy is contraindicated [197].

A pregnant woman in the household is not a contraindication for varicella immunization of a child or other contacts. Transmission of vaccine virus from an immunocompetent vaccine recipient to a non-immune individual has been reported only rarely, and only when a vaccine-associated rash develops in the vaccine recipient [2, 3].

For non-immune mothers exposed to varicella infection in the first or second trimester, varicella-zoster immune globulin (VZIG) should be administered within 72 hours to 10 days of exposure. If VZIG is not available, a dose of IVIG at 400 mg/kg can be used instead [2, 3]. If results of serologic testing are not available within this time frame, then postexposure prophylaxis should be offered. Patients need careful follow-up for signs of infection despite passive immunization.

In the absence of both VZIG and IVIG, prophylaxis with acyclovir (80 mg/kg/day, administered four times per day for 7 days; maximum dose, 800 mg, four times per day) may be administered beginning 7–10 days after exposure for people without evidence of immunity and with contraindications for varicella vaccination [2, 3], although published data on the benefit of acyclovir as post-exposure prophylaxis among immunocompromised people is limited.

1.6.7 Prognosis

Overall, the prognosis of infants born with CVS is poor. Mortality rates are estimated to be 30%. Deaths generally occur within the first few months of life resulting from intractable gastroesophageal reflux, severe recurrent aspiration pneumonia and respiratory failure. However, there are some reports of more favorable outcomes [120, 198]. The prognosis of neonatal varicella is variable, ranging from a mild illness to a disseminated infection, the mortality rate for neonatal varicella was 31% prior to the availability of VZIG, acyclovir, and neonatal intensive care, and is about 7% with optimal management in the current NICU care setting [199].

1.7   Mother-to-Child HIV Transmission

Perinatal transmission is the most common source of HIV infection among infants and children [200]. Remarkable progress has been made in prevention of perinatal transmission of HIV during the last decade in developed nations [201]. Rates of perinatal HIV transmission in the United States and Europe have decreased to 2%

or less because of widespread implementation of universal antenatal HIV testing, combination antiretroviral treatment during pregnancy, elective cesarean section, and avoidance of breast-feeding [202].

Availability of highly active antiretroviral therapy (HAART) has led to improved survival of HIV-infected children into adolescence and adulthood changing most HIV infections into a treatable chronic disease rather than a fatal disease [203].

Prevention of mother-to-child transmission of HIV is a major public health challenge in resource-limited countries [204]. Although several effective, simple, and less expensive prophylactic antiretroviral regimens are available to prevent perinatal HIV transmission, less than 10% of HIV-infected pregnant women have access to these preventive interventions in some countries in the developing world [205].

1.7.1 Transmission

Mother-to-child transmission of HIV-1 can occur in-utero, during labor and delivery, or postnatally through breast-feeding. Data suggest that most children are infected during the immediate peripartum period [206].

In the United States, the transmission rate without intervention is estimated to be 25–30%; in Europe, it is lower at 15–20% [207]. A transmission rate of 25–45% was previously observed among breast-feeding populations in Africa [208]. These variations in transmission rates likely reflect differences in infant feeding patterns, maternal and obstetric risk factors, viral factors, and differences in methodology among studies. Maternal disease status, especially a high viral load or a CD4$^+$ count less than 200 cells/mm^3, is highly correlated, however, with the risk for vertical transmission [209]. Obstetric risk factors associated with increased risk of transmission include vaginal delivery, rupture of membranes for more than 4 hours, chorioamnionitis, and invasive obstetric procedures [210]. Premature infants born to HIV-infected women have a higher rate of perinatal HIV infection than full-term infants [211].

1.7.2 Clinical Manifestations

The clinical manifestations of HIV infection in infants are highly variable and often nonspecific [205]. Infants with perinatally acquired HIV infection are often asymptomatic, and physical examination is usually normal in the neonatal period. A distinctive dysmorphic syndrome has not been confirmed [212, 213].

Growth delay is an early and frequent finding of untreated perinatal HIV infection, and linear growth is most severely affected in children with high viral loads [214]. Common clinical features seen during the first year of life include lymphadenopathy and hepatosplenomegaly [205].

Other manifestations noted during the course of HIV infection in children are failure to thrive, unexplained persistent fevers, developmental delay, encephalopathy, recurrent and chronic otitis media and sinusitis, recurrent invasive bacterial infections, opportunistic infections, chronic diarrhea, presence and persistence of oral candidiasis, parotitis, cardiomyopathy, nephropathy, and many non-specific cutaneous manifestations [215].

Infections in HIV-infected infants, not receiving antiretroviral therapy, can be serious or life-threatening. The difficulty in treating these infectious episodes, their chronicity, and their tendency to recur, distinguish them from the normal infections of early infancy. It is helpful to document each episode and to evaluate the course and frequency of their recurrences. With early diagnosis and access to antiretroviral treatment, opportunistic infections rarely develop in HIV-infected children [205].

Infections in HIV-infected newborns have the same pattern as seen commonly in the neonatal period. Oral candidiasis is common even in healthy, non-HIV-infected newborns and infants [205]. A syndrome of very-late-onset group B streptococcal disease (at 3.5–5 months of age) has been described in HIV-infected children [216]. Other infections include neonatal gonococcal disease or congenital syphilis [217].

1.7.3 Prevention of Mother-to-Child Transmission

1.7.3.1 Preconception Counseling for HIV-Infected Women
All women of childbearing age should be offered comprehensive family planning and the opportunity to receive preconception counseling and care as a component of routine primary medical care aiming to identify risk factors for adverse outcomes, provide education and counseling targeted to patients' individual needs, and treating or stabilizing medical conditions to optimize maternal and fetal outcomes [218]. Women living with HIV can continue to use some forms of contraceptive methods [219] taking into account drug interactions between hormonal contraceptives and antiretroviral therapy (ART).

Before conception is attempted by HIV-serodiscordant couples, the HIV-infected partner should be receiving ART and demonstrate sustained suppression of plasma viral load below the limits of detection. Use of ART reduces, but may not completely eliminate, the risk of HIV sexual transmission in couples who have decided to conceive through unprotected intercourse [220]. Starting ART before conception in HIV-infected women may also reduce the risk of perinatal transmission but not complete elimination of the risk of perinatal transmission [221].

1.7.3.2 Antepartum Care for HIV-Infected Women
ART for prevention of perinatal transmission of HIV are recommended for all pregnant women, regardless of CD4+ cell counts and HIV RNA levels. ART decreases maternal viral load in blood and genital secretions [222]. Clinicians should be aware of a possible small increased risk of preterm birth in pregnant women receiving protease-inhibitor (PI)-based combination antiretroviral therapy. However, given the clear benefits of such regimens for both a woman's health and

the prevention of perinatal transmission, PIs should not be withheld for fear of altering pregnancy outcome [223].

The WHO recommended first-line ART regimen in pregnant and breastfeeding women is the same as recommended in non-pregnant adults: tenofovir plus lamivudine or emtricitabine plus efavirenz administered as a fixed dose, once-daily combination regimen. Continued monitoring for adverse effects in HIV-infected pregnant women and birth defects in their infants is necessary to assure both short-term and long-term safety of ART regimen.

1.7.3.3 Intrapartum Care for HIV-Infected Women

ART is recommended for all woman who are HIV-infected. Some guidelines recommend scheduled cesarean delivery at 38 weeks' gestation (compared to 39 weeks for most indications) is recommended for women who have HIV RNA >1000 copies/mL near delivery [223]. Elective cesarean section is associated with lower rates of mother-to-child transmission among women who have received either no antiretroviral drugs or zidovudine alone [224] and is thus recommended for women who have not achieved viral suppression (HIV viral load >1000 copies/mL) in resource-rich regions (Panel on Treatment of HIV-Infected Pregnant Women and Prevention of Perinatal Transmission, 2016). However, this recommendation is not recommended in some regions of the world.

1.7.3.4 Postpartum Care

All current guidelines recommend lifelong ART for all HIV-infected women. Breastfeeding is not recommended for HIV-infected women in the United States [223].

All infants born to HIV-infected mothers should receive post-exposure antiretroviral prophylaxis [225]. Poor adherence to the antiretroviral agents can result in multi-class resistance in both the mother and infant; thus continued counseling and adherence evaluation is crucial for prevention of mother-to-child transmission.

References

1. Karwowski MP, Nelson JM, Staples JE, et al. Zika virus disease: a CDC update for pediatric health care providers. Pediatrics. 2016;137(5):e2016062.
2. Centers for Disease Control and Prevention (CDC). CDC Health Information for International Travel 2016. New York: Oxford University Press; 2016.
3. Centers for Disease Control and Prevention (CDC). Zika virus: symptoms, diagnosis, & treatment. 2016. http://www.cdc.gov/zika/symptoms/index.html. Accessed 3 Aug 2016.
4. Colt S, Garcia-Casal MN, Peña-Rosas JP, et al. Transmission of Zika virus through breast milk and other breastfeeding-related bodily-fluids: a systematic review. Bull World Health Organ; 2016. E-pub: 2 May 2016.
5. Bonaldo MC, Ribeiro IP, Lima NS, et al. Isolation of infective Zika Virus from urine and saliva of patients in Brazil. PLoS Negl Trop Dis. 2016;10(6):e0004816.
6. Petersen EE, Staples JE, Meaney-Delman D, et al. Interim guidelines for pregnant women during a Zika Virus outbreak—United States, 2016. MMWR Morb Mortal Wkly Rep. 2016;65(2):30.
7. Petersen LR, Jamieson DJ, Powers AM, et al. Zika virus. N Engl J Med. 2016;374:1552–156.

8. Cugola FR, Fernandes IR, Russo FB, et al. The Brazilian Zika virus strain causes birth defects in experimental models. Nature. 2016;534(7606):267–71.
9. Vouga M, Baud D. Imaging of congenital Zika virus infection: the route to identification of prognostic factors. Prenat Diagn. 2016;36(9):799.
10. Moron AF, Cavalheiro S, Milani HJF, et al. Microcephaly associated with maternal Zika virus infection. BJOG. 2016;123:1265–9.
11. van der Linden V, Filho ELR, Lins OG, et al. Congenital Zika syndrome with arthrogryposis: retrospective case series study. BMJ. 2016;354:i3899.
12. Driggers RW, Ho CY, Korhonen EM, et al. Zika virus infection with prolonged maternal viremia and fetal brain abnormalities. N Engl J Med. 2016;374:2142.
13. Rasmussen SA, Denise JJ, Honein MA, et al. Zika virus and birth defects—reviewing the evidence for causality. N Engl J Med. 2016;374:1981–7.
14. von der Hagen M, Pivarcsi M, Liebe J, et al. Diagnostic approach to microcephaly in childhood: a two center study and review of the literature. Dev Med Child Neurol. 2014;56(8): 732–41.
15. de Paula FB, de Oliveira DJ, Prazeres J, et al. Ocular findings in infants with microcephaly associated with presumed Zika virus congenital infection in Salvador, Brazil. JAMA Ophthalmol. 2016;134(5):529–35.
16. De Noronha L, Zanluca C, Azevedo MLV, et al. Zika virus damages the human placental barrier and presents marked fetal neurotropism. Mem Inst Oswaldo Cruz. 2016;111(5):287–93.
17. World Health Organization (WHO). Zika virus and complications: questions and answers. 2016. http://www.who.int/features/qa/zika/en/. Accessed 9 Aug 2016.
18. Shrestha P, Horby P, Carson G. Non-vector transmission of flaviviruses, with implications for the Zika virus. Bull World Health Organ; 2016. E-pub: 6 Jul 2016.
19. Fleming-Dutra KE, Nelson JM, Fischer M, et al. Update: interim guidelines for health care providers caring for infants and children with possible Zika virus infection—United States, February 2016. MMWR Morb Mortal Wkly Rep. 2016;65:182–7.
20. Food and Drug Administration (FDA). Recommendations for donor screening, deferral, and product management to reduce the risk of transfusion transmission of Zika virus. Guidance for Industry. 2016. http://www.fda.gov/downloads/BiologicsBloodVaccines/GuidanceComplianceRegulatoryInformation/Guidances/Blood/UCM486360.pdf. Accessed 10 Sept 2016.
21. Vouga M, Musso D, Van Mieghem T, et al. CDC guidelines for pregnant women during the Zika virus outbreak. Lancet. 2016;387:843.
22. Reef SE, Plotkin S, Cordero JF, et al. Preparing for elimination of congenital Rubella syndrome (CRS): summary of a workshop on CRS elimination in the United States. Clin Infect Dis. 2000;31(1):85.
23. Reef SE, Frey TK, Theall K, et al. The changing epidemiology of rubella in the 1990s: on the verge of elimination and new challenges for control and prevention. JAMA. 2002; 287(4):464.
24. Sheridan E, Aitken C, Jeffries D, et al. Congenital rubella syndrome: a risk in immigrant populations. Lancet. 2002;359(9307):674.
25. McElroy R, Laskin M, Jiang D, et al. Rates of rubella immunity among immigrant and non-immigrant pregnant women. J Obstet Gynaecol Can. 2009;31(5):409.
26. Cherry JD, Adachi K. Rubella virus. In: Cherry JD, Harrison GJ, Kaplan SL, et al, editors. Feigin and Cherry's textbook of pediatric infectious diseases. 7th ed. Philadelphia: Elsevier Saunders; 2014. p. 2195.
27. Morgan-Capner P, Miller E, Vurdien JE, et al. Outcome of pregnancy after maternal reinfection with rubella. CDR (Lond Engl Rev). 1991;1(6):R57.
28. Webster WS. Teratogen update: congenital rubella. Teratology. 1998;58(1):13.
29. Miller E, Cradock-Watson JE, Pollock TM. Consequences of confirmed maternal rubella at successive stages of pregnancy. Lancet. 1982;2(8302):781.
30. Enders G, Nickerl-Pacher U, Miller E, et al. Outcome of confirmed periconceptional maternal rubella. Lancet. 1988;1(8600):1445.

31. Naeye RL, Blanc W. Pathogenesis of congenital rubella. JAMA. 1965;194(12):1277.
32. Rawls WE. Congenital rubella: the significance of virus persistence. Prog Med Virol. 1968;10:238.
33. Lee JY, Bowden DS. Rubella virus replication and links to teratogenicity. Clin Microbiol Rev. 2000;13(4):571.
34. Plotkin SA, Vaheri A. Human fibroblasts infected with rubella virus produce a growth inhibitor. Science. 1967;156(3775):659.
35. Duncan R, Muller J, Lee N, et al. Rubella virus-induced apoptosis varies among cell lines and is modulated by Bcl-XL and caspase inhibitors. Virology. 1999;255(1):117.
36. Adamo P, Asís L, Silveyra P, et al. Rubella virus does not induce apoptosis in primary human embryo fibroblast cultures: a possible way of viral persistence in congenital infection. Viral Immunol. 2004;17(1):87.
37. Banatvala JE, Brown DW. Rubella. Lancet. 2004;363(9415):1127.
38. Centers for Disease Control and Prevention (CDC). Control and prevention of rubella: evaluation and management of suspected outbreaks, rubella in pregnant women, and surveillance for congenital rubella syndrome. MMWR Recomm Rep. 2001;50(RR12):1–23. https://www.cdc.gov/mmwr/preview/mmwrhtml/rr5012a1.htm. Accessed 9 Sept 2016.
39. Edlich RF, Winters KL, Long 3rd WB, et al. Rubella and congenital rubella (German measles). J Long Term Eff Med Implants. 2005;15(3):319–28.
40. Gabbe SG, Niebyl JR, Simpson JL. Obstetrics-normal and problem pregnancies. 4th ed. New York: Churchill Livingstone; 2002. p. 1328–30.
41. Johnson RE, Hall AP. Rubella arthritis. N Engl J Med. 1958;258:743–5.
42. Plotkin, SA, Reef, SE, Cooper, LZ, et al. Rubella. In: Remington JS, Klein JO, Wilson CB, et al, editors. Infectious diseases of the fetus and newborn infant. 7th ed. Philadelphia: Elsevier Saunders; 2011. p. 861.
43. Oster ME, Riehle-Colarusso T, Correa A, et al. An update on cardiovascular malformations in congenital rubella syndrome. Birth Defects Res A Clin Mol Teratol. 2010;88(1):1.
44. Givens KT, Lee DA, Jones T, et al. Congenital rubella syndrome: ophthalmic manifestations and associated systemic disorders. Br J Ophthalmol. 1993;77(6):358–63.
45. Viskari H, Paronen J, Keskinen P, et al. Humoral beta-cell autoimmunity is rare in patients with the congenital rubella syndrome. Clin Exp Immunol. 2003;133(3):378.
46. Dontigny L, Arsenault M, Martel M, et al. Rubella in pregnancy. J Obstet Gynaecol Can. 2008;30(2):152–8.
47. Centers for Disease Control (CDC). Control and prevention of rubella: evaluation and management of suspected outbreaks, rubella in pregnant women, and surveillance for congenital rubella syndrome. MMWR Recomm Rep. 2001;50(RR12);1–23.
48. Centers for Disease Control and Prevention (CDC). Prevention of measles, rubella, congenital rubella syndrome, and mumps, 2013: summary recommendations of the Advisory Committee on Immunization Practices (ACIP). MMWR. 2013;62(RR-4):13.
49. Bart SW, Stetler HC, Preblud SR, et al. Fetal risk associated with rubella vaccine: an update. Rev Infect Dis. 1985;7 Suppl 1:S95–102.
50. Maldonado YA. Rubella virus. In: Long SS, Pickering LK, Prober CG, editors. Principles and practice of pediatric infectious diseases. 4th ed. Edinburgh: Elsevier Saunders; 2012. p. 1112.
51. American Academy of Pediatrics (AAP). Rubella. In: Kimberlin DW, Brady MT, Jackson MA, Long SS, editors. Red Book: 2015 Report of the Committee on Infectious Diseases. 30th ed. Elk Grove Village: American Academy of Pediatrics; 2015. p. 688.
52. Cooper LZ. The history and medical consequences of rubella. Rev Infect Dis. 1985;7 Suppl 1:S2.
53. Forrest JM, Turnbull FM, Sholler GF, et al. Gregg's congenital rubella patients 60 years later. Med J Aust. 2002;177(11–12):664.
54. McIntosh ED, Menser MA. A fifty-year follow-up of congenital rubella. Lancet. 1992;340(8816):414.
55. Menser MA, Dods L, Harley JD. A twenty-five-year follow-up of congenital rubella. Lancet. 1967;2(7530):1347.

56. Harrison GJ. Cytomegalovirus. In: Cherry JD, Harrison GJ, Kaplan SL, et al, editors. Feigin and Cherry's textbook of pediatric infectious diseases, 7th ed. Philadelphia: Elsevier Saunders; 2014. p. 1969.

57. Swanson EC, Schleiss MR. Congenital cytomegalovirus infection: new prospects for prevention and therapy: for pediatric clinics of North America: advances in evaluation, diagnosis and treatment of pediatric infectious disease. Pediatr Clin North Am. 2013;60(2):335–49.

58. Johnson J, Anderson B. Screening, prevention, and treatment of congenital cytomegalovirus. Obstet Gynecol Clin North Am. 2014;41:593–9.

59. Lantos PM, Permar SR, Hoffman K, et al. The excess burden of cytomegalovirus in African American communities: a geospatial analysis. Open Forum Infect Dis. 2015;20(2):ofv180.

60. Boppana SB, Rivera LB, Fowler KB, et al. Intrauterine transmission of cytomegalovirus to infants of women with preconceptional immunity. N Engl J Med. 2001;344:1366–71.

61. Lazzarotto T, Guerra B, Gabrielli L, et al. Update on the prevention, diagnosis and management of cytomegalovirus infection during pregnancy. Clin Microbiol Infect. 2011;17: 1285–93.

62. American Academy of Pediatrics (AAP). Cytomegalovirus infection. In: Kimberlin DW, editor. Red Book: 2015 Report of the Committee on Infectious Diseases. 30th ed. Elk Grove Village: American Academy of Pediatrics; 2015. p. 317–2.

63. Kenneson A, Cannon MJ. Review and meta-analysis of the epidemiology of congenital cytomegalovirus (CMV) infection. Rev Med Virol. 2007;17(4):253.

64. Goderis J, De Leenheer E, Smets K, et al. Hearing loss and congenital CMV infection: a systematic review. Pediatrics. 2014;134(5):972.

65. Noyola DE, Demmler GJ, Williamson WD, et al. Cytomegalovirus urinary excretion and long term outcome in children with congenital cytomegalovirus infection. Congenital CMV Longitudinal Study Group. Pediatr Infect Dis J. 2000;19:505–10.

66. Lazzarotto T, Guerra B, Lanari M, et al. New advances in the diagnosis of congenital cytomegalovirus infection. J Clin Virol. 2008;41:192–7.

67. van Zuylen WJ, Ford CE, Wong DD, et al. Human cytomegalovirus modulates expression of noncanonical Wnt receptor ROR2 to alter trophoblast migration. J Virol. 2015;90:1108–15.

68. Cheeran MC, Lokensgard JR, Schleiss MR. Neuropathogenesis of congenital cytomegalovirus infection: disease mechanisms and prospects for intervention. Clin Microbiol Rev. 2009;22: 99–126.

69. Picone O, Vauloup-Fellous C, Cordier AG, et al. A series of 238 cytomegalovirus primary infections during pregnancy: description and outcome. Prenat Diagn. 2013;33:751–8.

70. Bany-Mohammed F. Cytomegalovirus. In: Gomella TL, Cunningham MD, Eyal FG, Tuttle D, editors. Neonatology: management, procedures, on-call problems, diseases and drugs. 7th ed. New York: McGraw-Hill; 2013. p. 615–19.

71. Guerra B, Simonazzi G, Puccetti C, et al. Ultrasound prediction of symptomatic congenital cytomegalovirus infection. Am J Obstet Gynecol. 2008;198(380):e1–7.

72. Picone O, Teissier N, Cordier AG, et al. Detailed in utero ultrasound description of 30 cases of congenital cytomegalovirus infection. Prenat Diagn. 2014;34:518–24.

73. Capretti MG, Lanari M, Tani G, et al. Role of cerebral ultrasound and magnetic resonance imaging in newborns with congenital cytomegalovirus infection. Brain Dev. 2014;36: 203–11.

74. Istas AS, Demmler GJ, Dobbins JG, et al. Surveillance for congenital cytomegalovirus disease: a report from the National Congenital Cytomegalovirus Disease Registry. Clin Infect Dis. 1995;20(3):665.

75. Boppana SB, Ross SA, Fowler KB. Congenital cytomegalovirus infection: clinical outcome. Clin Infect Dis. 2013;57 Suppl 4:S178.

76. Ghekiere S, Allegaert K, Cossey V, et al. Ophthalmological findings in congenital cytomegalovirus infection: when to screen, when to treat? J Pediatr Ophthalmol Strabismus. 2012;49(5):274.

77. Britt W. Cytomegalovirus. In: Remington, Klein, editors. Infectious diseases of the fetus and neonate. 7th ed. Philadelphia: WB Saunders; 2011. p. 706.

78. Salisbury S, Embil JA. Graves disease following congenital cytomegalovirus infection. J Pediatr. 1978;92:954.
79. Mena W, Royal S, Pass RF, et al. Diabetes insipidus associated with symptomatic congenital cytomegalovirus infection. J Pediatr. 1993;122(6):911–3.
80. Besbas N, Bayrakci US, Kale G, et al. Cytomegalovirus-related congenital nephrotic syndrome with diffuse mesangial sclerosis. Pediatr Nephrol. 2006;5:740–2. Epub 2006 Mar 8.
81. Williamson WD, Demmler GJ, Percy AK, et al. Progressive hearing loss in infants with asymptomatic congenital cytomegalovirus infection. Pediatrics. 1992;90(6):862.
82. Goegebuer T, Van Meensel B, Beuselinck K, et al. Clinical predictive value of real-time PCR quantification of human cytomegalovirus DNA in amniotic fluid samples. J Clin Microbiol. 2009;47:660–5.
83. Society for Maternal-Fetal Medicine (SMFM), Hughes BL, Gyamfi-Bannerman C. Diagnosis and antenatal management of congenital cytomegalovirus infection. Am J Obstet Gynecol. 2016;214:B5–B11.
84. Adler SP. Primary maternal cytomegalovirus infection during pregnancy: do we have a treatment option? Clin Infect Dis. 2012;55:504–6.
85. Lazzarotto T, Spezzacatena P, Varani S, et al. Anticytomegalovirus (anti-CMV) immunoglobulin G avidity in identification of pregnant women at risk of transmitting congenital CMV infection. Clin Diagn Lab Immunol. 1999;6:127–9.
86. Carlson A, Norwitz ER, Stiller RJ. Cytomegalovirus infection in pregnancy: should all women be screened? Rev Obstet Gynecol. 2010;3(4):172–9.
87. Fink KR, Thapa MM, Ishak GE, et al. Neuroimaging of pediatric central nervous system cytomegalovirus infection. Radiographics. 2010;30(7):1779–96.
88. Barkovich AJ, Moore KR, Jones BV, et al. Diagnostic imaging: pediatric neuroradiology. Salt Lake City: AMIRSYS; 2007.
89. de Vries LS, Gunardi H, Barth PG, et al. The spectrum of cranial ultrasound and magnetic resonance imaging abnormalities in congenital cytomegalovirus infection. Neuropediatrics. 2004;35(2):113–9.
90. Zucca C, Binda S, Borgatti R, et al. Retrospective diagnosis of congenital cytomegalovirus infection and cortical maldevelopment. Neurology. 2003;61(5):710–2.
91. Noyola DE, Demmler GJ, Nelson CT, et al. Early predictors of neurodevelopmental outcome in symptomatic congenital cytomegalovirus infection. J Pediatr. 2001;138(3):325–31.
92. Fernández AA, Martín AP, Martínez MI, et al. Chronic fatigue syndrome. Summary of the consensus document [in Spanish]. Aten Primaria. 2009;41:e1–5.
93. Kimberlin DW, Whitley RJ. Antiviral therapy of HSV-1 and -2. In: Arvin A, Campadelli-Fiume G, Mocarski E, et al., editors. Human herpesviruses: biology, therapy, and immunoprophylaxis. Cambridge: Cambridge University Press; 2007.
94. Kimberlin DW, Lin CY, Sánchez PJ, et al. Effect of ganciclovir therapy on hearing in symptomatic congenital cytomegalovirus disease involving the central nervous system: a randomized, controlled trial. J Pediatr. 2003;143:16–25.
95. Kimberlin DW, Jester PM, Sánchez PJ, et al. Valganciclovir for symptomatic congenital cytomegalovirus disease. N Engl J Med. 2015;372:933–43.
96. Gwee A, Curtis N, Connell TG, et al. Ganciclovir for the treatment of congenital cytomegalovirus: what are the side effects? Pediatr Infect Dis J. 2014;33:115.
97. Ghandi RS, Fernandez-Alvarez JR, Rabe H. Management of congenital cytomegalovirus infection. Acta Paediatr. 2010;99:509–15.
98. Mareri A, Lasorella S, Iapadre G, et al. Anti-viral therapy for congenital cytomegalovirus infection: pharmacokinetics, efficacy and side effects. J Matern Fetal Neonatal Med. 2016;29:1657–64.
99. Leruez-Ville M, Ville Y. Optimum treatment of congenital cytomegalovirus infection. Expert Rev Anti Infect Ther. 2016;14(5):479–88.
100. Adler SP, Finney JW, Manganello AM, et al. Prevention of child-to-mother transmission of cytomegalovirus among pregnant women. J Pediatr. 2004;145:485–91.

101. Harvey J, Dennis CL. Hygiene interventions for prevention of cytomegalovirus infection among childbearing women: systematic review. J Adv Nurs. 2008;63:440–50.
102. Jückstock J, Rothenburger M, Friese K, et al. Passive immunization against congenital cytomegalovirus infection: current state of knowledge. Pharmacology. 2015;95(5–6):209–17.
103. Pass RF, Zhang C, Evans A, et al. Vaccine prevention of maternal cytomegalovirus infection. N Engl J Med. 2009;360(12):1191.
104. Jacquemard F, Yamamoto M, Costa JM, et al. Maternal administration of valacyclovir in symptomatic intrauterine cytomegalovirus infection. BJOG. 2007;114:1113–21.
105. American College of Obstetricians and Gynecologists (ACOG). Cytomegalovirus, parvovirus B19, varicella zoster, and toxoplasmosis in pregnancy. Practice bulletin no. 151. Obstet Gynecol. 2015;125:1510–25.
106. Rosenthal LS, Fowler KB, Boppanna SB, et al. Cytomegalovirus shedding and delayed sensorineural hearing loss: results from longitudinal follow-up of children with congenital infection. Pediatr Infect Dis J. 2009;28:515–20.
107. Stagno S, Whitley RJ. Herpesvirus infections of pregnancy. Part I: cytomegalovirus and Epstein-Barr virus infections. N Engl J Med. 1985;313:1270–4.
108. Fowler KB, McCollister FP, Dahle AJ, et al. Progressive and fluctuating sensorineural hearing loss in children with asymptomatic congenital cytomegalovirus infection. J Pediatr. 1997;130:624–30.
109. American Academy of Pediatrics (AAP), Joint Committee on Infant Hearing. Year 2007 position statement: principles and guidelines for early hearing detection and intervention programs. Pediatrics. 2007;120:898–921.
110. Townsend CL, Forsgren M, Ahlfors K, et al. Long-term outcomes of congenital cytomegalovirus infection in Sweden and the United Kingdom. Clin Infect Dis. 2013;56:1232–9.
111. Whitley RJ. Herpes simplex virus. In: Scheld MW, Whitley RJ, Marra CM, editors. Infections in the central nervous system. 3rd ed. Philadelphia: Lippincott Williams & Wilkins; 2004. p. 123–44.
112. James SH, Sheffield JS, Kimberlin DW. Mother-to-child transmission of herpes simplex virus. J Pediatric Infect Dis Soc. 2014;3(Suppl 1):S19–23.
113. James SH, Kimberlin DW. Neonatal herpes simplex virus infection. Infect Dis Clin North Am. 2015;29:391–400.
114. Sampath A, Maduro G, Schillinger JA. Infant deaths due to herpes simplex virus, congenital syphilis, and HIV in New York city. Pediatrics. 2016;137(4).
115. Caviness AC, Demmler GJ, Almendarez Y, et al. The prevalence of neonatal herpes simplex virus infection compared with serious bacterial illness in hospitalized neonates. J Pediatr 2008;153:164–9.
116. Pinninti SG, Kimberlin DW. Maternal and neonatal herpes simplex virus infections. Am J Perinatol. 2013;30:113–9.
117. ACOG Practice Bulletin. Clinical management guidelines for obstetrician-gynecologists. No. 82 June 2007. Management of herpes in pregnancy. Obstet Gynecol. 2007;109:1489–98.
118. Kimberlin DW. Herpes simplex virus infections of the newborn. Semin Perinatol. 2007;31:19–25.
119. Brown ZA, Wald A, Morrow RA, et al. Effect of serologic status and cesarean delivery on transmission rates of herpes simplex virus from mother to infant. JAMA. 2003;289:203–9.
120. Sauerbrei A, Wutzler P. Herpes simplex and varicella-zoster virus infections during pregnancy: current concepts of prevention, diagnosis and therapy. Part 1: herpes simplex virus infections. Med Microbiol Immunol. 2007;196:89–94.
121. Arvin AM, Whitley RJ, Gutierrez KM. Herpes simplex virus infections. In: Infectious diseases of the foetus and newborn infant. Philadelphia: Elsevier Saunders; 2006. p. 845–65.
122. Kimberlin DW, Lin C-Y, Jacobs RF. Natural history of neonatal herpes simplex virus infections in the acyclovir era. Pediatrics. 2001;108:223–9.
123. Kimberlin DW, Lin CY, Jacobs RF, et al. Safety and efficacy of high-dose intravenous acyclovir in the management of neonatal herpes simplex virus infections. Pediatrics. 2001;108:230–138.

124. Whitley RJ, Gnann JW Jr. Herpes simplex virus. In: Tyring SK, Yen-Moore A, editors. Mucocutaneous manifestations of viral diseases. London: Informa Health Care; 2002. p. 69–117.

125. Curfman AL, Glissmeyer EW, Ahmad FA, et al. Initial presentation of neonatal herpes simplex virus infection. J Pediatr. 2016;172:121–6.

126. Le Doare K, Menson E, Patel D, et al. Fifteen-minute consultation: managing neonatal and childhood herpes encephalitis. Arch Dis Child Educ Pract Ed. 2015;100:58–63.

127. Kotzbauer D, Frank G, Dong W, et al. Clinical and laboratory characteristics of disseminated herpes simplex virus infection in neonates. Hosp Pediatr. 2014;4:167–71.

128. Kimberlin DW, Baley J, Committee on Infectious Diseases, Committee on Fetus and Newborn. Guidance on management of asymptomatic neonates born to women with active genital herpes lesions. Pediatrics. 2013;131(2):383–6.

129. American Academy of Pediatrics (AAP). Herpes simplex. In: Kimberlin DW, editor. Red book: 2015 report of the committee on infectious diseases. 30th ed. Elk Grove Village: American Academy of Pediatrics; 2015. p 432-43.

130. Frenkel LM. Challenges in the diagnosis and management of neonatal herpes simplex virus encephalitis. Pediatrics. 2005;115:795–7.

131. Melvin AJ, Mohan KM, Schiffer JT, et al. Plasma and cerebrospinal fluid herpes simplex virus levels at diagnosis and outcome of neonatal infection. J Pediatr. 2015;166:827–33.

132. Malm G, Forsgren M. Neonatal herpes simplex virus infections: HSV DNA in cerebrospinal fluid and serum. Arch Dis Child Fetal Neonatal Ed. 1999;81(1):F24–9.

133. Kimberlin DW, Whitley RJ, Wan W, et al. Oral acyclovir suppression and neurodevelopment after neonatal herpes. N Engl J Med. 2011;365:1284–92.

134. Egawa H, Inomata Y, Nakayama S, et al. Fulminant hepatic failure secondary to herpes simplex virus infection in a neonate: a case report of successful treatment with liver transplantation and perioperative acyclovir. Liver Transpl Surg. 1998;4:513–5.

135. Maeba S, Hasegawa S, Shimomura M, et al. Successful treatment of corticosteroid with antiviral therapy for a neonatal liver failure with disseminated herpes simplex virus infection. AJP Rep. 2015;5:e89–92.

136. Sheffield JS, Hollier LM, Hill JB, et al. Acyclovir prophylaxis to prevent herpes simplex virus recurrence at delivery: a systematic review. Obstet Gynecol. 2003;102:1396–403.

137. Pinninti SG, Angara R, Feja KN, et al. Neonatal herpes disease following maternal antenatal antiviral suppressive therapy: a multicenter case series. J Pediatr. 2012;161:134–8.e1–3.

138. ACOG. Management of herpes in pregnancy. Obstet Gynecol. 2007;109:1489–98.

139. Belshe RB, Leone PA, Bernstein DI, et al. Efficacy results of a trial of a herpes simplex vaccine. N Engl J Med. 2012;366:34–43.

140. Whitley R, Arvin A, Prober C, et al. Predictors of morbidity and mortality in neonates with herpes simplex virus infections. The National Institute of Allergy and Infectious Diseases Collaborative Antiviral Study Group. N Engl J Med. 1991;324:450–4.

141. Whitley RJ, Corey L, Arvin A, et al. Changing presentation of herpes simplex virus infection in neonates. J Infect Dis. 1988;158:109–16.

142. Fonseca-Aten M, Messina AF, Jafri HS, et al. Herpes simplex virus encephalitis during suppressive therapy with acyclovir in a premature infant. Pediatrics. 2005;115:804–9.

143. Brown KE, Anderson SM, Young NS. Erythrocyte P antigen: cellular receptor for B19 parvovirus. Science. 1993;262(5130):114.

144. Gratacós E, Torres PJ, Vidal J, et al. The incidence of human parvovirus B19 infection during pregnancy and its impact on perinatal outcome. J Infect Dis. 1995;171(5):1360.

145. Markenson GR, Yancey MK. Parvovirus B19 infections in pregnancy. Semin Perinatol. 1998;22:309–17.

146. Adler S, Koch WC. Human parvovirus B19. In: Remington JS, Klein JO, editors. Infectious diseases of the fetus and newborn infant. 7th ed. Philadelphia: Saunders; 2011. p. 845–5.

147. Crane J, Mundle W, Boucoiran I, et al. Parvovirus B19 infection in pregnancy. J Obstet Gynaecol Can. 2014;36(12):1107–16.

148. Harger JH, Adler SP, Koch WC, et al. Prospective evaluation of 618 pregnant women exposed to parvovirus B19: risks and symptoms. Obstet Gynecol. 1998;91:413–20.
149. Morey AL, Ferguson DJ, Fleming KA. Ultrastructural features of fetal erythroid precursors infected with parvovirus B19 in vitro: evidence of cell death by apoptosis. J Pathol. 1993;169(2):213.
150. Rodis JF. Parvovirus infection. Clin Obstet Gynecol. 1999;42:107–20; quiz 174–5.
151. Watt AP, Brown M, Pathiraja M, et al. The lack of routine surveillance of parvovirus B19 infection in pregnancy prevents an accurate understanding of this regular cause of fetal loss and the risks posed by occupational exposure. J Med Microbiol. 2013;62(Pt 1):86–92.
152. Jordan JA. Identification of human parvovirus B19 infection in idiopathic nonimmune hydrops fetalis. Am J Obstet Gynecol. 1996;174(1 Pt 1):37.
153. Puccetti C, Contoli M, Bonvicini F, et al. Parvovirus B19 in pregnancy: possible consequences of vertical transmission. Prenat Diagn. 2012;32(9):897–902. Epub 2012 Jul 9.
154. Rodis JF, Borgida AF, Wilson M, et al. Management of parvovirus infection in pregnancy and outcomes of hydrops: a survey of members of the Society of Perinatal Obstetricians. Am J Obstet Gynecol. 1998;179(4):985.
155. Marton T, Martin WL, Whittle MJ. Hydrops fetalis and neonatal death from human parvovirus B19: an unusual complication. Prenat Diagn. 2005;25(7):543.
156. Segata M, Chaoui R, Khalek N, et al. Fetal thrombocytopenia secondary to parvovirus infection. Am J Obstet Gynecol. 2007;196(1):61.e1.
157. Dijkmans AC, de Jong EP, Dijkmans BA, et al. Parvovirus B19 in pregnancy: prenatal diagnosis and management of fetal complications. Curr Opin Obstet Gynecol. 2012;24:95–101.
158. Parilla BV, Tamura RK, Ginsberg NA. Association of parvovirus infection with isolated fetal effusions. Am J Perinatol. 1997;14(6):357.
159. Katz VL, McCoy MC, Kuller JA, et al. An association between fetal parvovirus B19 infection and fetal anomalies: a report of two cases. Am J Perinatol. 1996;13(1):43.
160. Tiessen RG, van Elsacker-Niele AM, Vermeij-Keers C, et al. A fetus with a parvovirus B19 infection and congenital anomalies. Prenat Diagn. 1994;14(3):173.
161. Miller E, Fairley CK, Cohen BJ, et al. Immediate and long term outcome of human parvovirus B19 infection in pregnancy. Br J Obstet Gynaecol. 1998;105(2):174.
162. Ergaz Z, Ornoy A. Parvovirus B19 in pregnancy. Reprod Toxicol. 2006;21(4):421.
163. Rodis JF, Hovick Jr TJ, Quinn DL, et al. Human parvovirus infection in pregnancy. Obstet Gynecol. 1988;72:733–8.
164. Enders M, Schalasta G, Baisch C, et al. Human parvovirus B19 infection during pregnancy—value of modern molecular and serological diagnostics. J Clin Virol. 2006;35:400–6.
165. Levy R, Weissman A, Blomberg G, et al. Infection by parvovirus B19 during pregnancy: a review. Obstet Gynecol Surv. 1997;52:254–9.
166. Delle Chiaie L, Buck G, Grab D, et al. Prediction of fetal anemia with Doppler measurement of the middle cerebral artery peak systolic velocity in pregnancies complicated by maternal blood group alloimmunization or parvovirus B19 infection. Ultrasound Obstet Gynecol. 2001;18:232–6.
167. Bernstein IM, Capeless EL. Elevated maternal serum alpha-fetoprotein and hydrops fetalis in association with fetal parvovirus B-19 infection. Obstet Gynecol. 1989;74(3 Pt 2):456–7.
168. Pasman SA, Claes L, Lewi L, et al. Intrauterine transfusion for fetal anemia due to red blood cell alloimmunization: 14 years experience in Leuven. Facts Views Vis Obgyn. 2015;7(2):129–36.
169. Canlorbe G, Macé G, Cortey A, et al. Management of very early fetal anemia resulting from red-cell alloimmunization before 20 weeks of gestation. Obstet Gynecol. 2011;118:1323.
170. Kumar A, Dutta AK. Intrauterine and perinatal infections—an overview. In: Sachdeva A, Dutta A, editors. Advances in pediatrics. 2nd ed. London: JP Medical Ltd; 2012. p. 149–63.
171. Young NS, Brown KE. Parvovirus B19. N Engl J Med. 2004;350(6):586.
172. Seng C, Watkins P, Morse D, et al. Parvovirus B19 outbreak on an adult ward. Epidemiol Infect. 1994;113(2):345.

173. Centers for Disease Control (CDC). Risks associated with human parvovirus B19 infection. MMWR Morb Mortal Wkly Rep. 1989;38(6):81.
174. Bernstein DI, El Sahly HM, Keitel WA, et al. Safety and immunogenicity of a candidate parvovirus B19 vaccine. Vaccine. 2011;43:7357–63. Epub 2011 Jul 30.
175. American Academy of Pediatrics (AAP). Varicella-Zoster virus infections. In: Kimberlin DW, Brady MT, Jackson MA, Long SS, editors. Red Book: 2015 Report of the Committee on Infectious Diseases. 30th ed. Elk Grove Village: American Academy of Pediatrics; 2015. p. 847.
176. Birthistle K, Carrington D. Foetal varicella syndrome—a reappraisal of the literature. A review prepared for the UK Advisory Group on Chickenpox on behalf of the British Society for the Study of Infection. J Infect. 1998;36:25–9.
177. Cohen A, Moschopoulos P, Stiehm RE, et al. Congenital varicella syndrome: the evidence for secondary prevention with varicella-zoster immune globulin. CMAJ. 2011;183(2):204.
178. Pastuszak AL, Levy M, Schick B, et al. Outcome after maternal varicella infection in the first 20 weeks of pregnancy. N Engl J Med. 1994;330(13):901.
179. Grose C. Variation on a theme by Fenner: the pathogenesis of chickenpox. Pediatrics. 1981;68:735–7.
180. Sauerbrei A, Wutzler P. The congenital varicella syndrome. J Perinatol. 2000;20:548–54.
181. Saji F, Samejima Y, Kamiura S, et al. Dynamics of immunoglobulins at the feto-maternal interface. Rev Reprod. 1999;4:81–9.
182. Higa K, Dan K, Manabe H. Varicella-zoster virus infections during pregnancy: hypothesis concerning the mechanisms of congenital malformations. Obstet Gynecol. 1987;69:214–22.
183. Grose C. Congenital varicella-zoster virus infection and the failure to establish virus-specific cell-mediated immunity. Mol Biol Med. 1989;6:453–62.
184. Gershon AA. Chickenpox, measles, and mumps. In: Remington, Klein, editors. Infectious diseases of the fetus and neonate. 7th ed. Philadelphia: WB Saunders; 2011. p. 661.
185. Enders G, Miller E, Cradock-Watson J, et al. Consequences of varicella and herpes zoster in pregnancy: prospective study of 1739 cases. Lancet. 1994;343:1548–51.
186. Sauerbrei A, Muller D, Eichhorn U, et al. Detection of varicella zoster virus in congenital varicella syndrome: a case report. Obstet Gynecol. 1996;88:687–9.
187. Enders G, Miller E. Varicella and herpes zoster in pregnancy and the newborn. In: Arvin AM, Gershon AA, VZV Research Foundation, editors. Varicella-zoster virus: virology and clinical management. Cambridge: Cambridge University Press; 2000. p. 317–47.
188. De Paschale M, Clerici P. Microbiology laboratory and the management of mother-child varicella-zoster virus infection. World J Virol. 2016;5:97–124.
189. Arvin AM, Koropchak CM, Wittek AE. Immunologic evidence of reinfection with varicella-zoster virus. J Infect Dis. 1983;148:200–5.
190. Gershon AA, Steinberg SP, Borkowsky W, et al. IgM to varicella-zoster virus: demonstration in patients with and without clinical zoster. Pediatr Infect Dis. 1982;1:164–7.
191. Kerkering KW. Abnormal cry and intracranial calcifications: clues to the diagnosis of fetal varicella-zoster syndrome. J Perinatol. 2001;21:131–5.
192. Stone KM, Reiff-Eldridge R, White AD, et al. Pregnancy outcomes following systemic prenatal acyclovir exposure: conclusions from the international acyclovir pregnancy registry, 1984-1999. Birth Defects Res A Clin Mol Teratol. 2004;70(4):201.
193. Smego Jr RA, Asperilla MO. Use of acyclovir for varicella pneumonia during pregnancy. Obstet Gynecol. 1991;78(6):1112.
194. Reiff-Eldridge R, Heffner CR, Ephross SA, et al. Monitoring pregnancy outcomes after prenatal drug exposure through prospective pregnancy registries: a pharmaceutical company commitment. Am J Obstet Gynecol. 2000;182(1 Pt 1):159.
195. Baley JE, Gonzalez BE. Perinatal viral infections. In: Fanaroff, Martin, editors. Neonatal-perinatal medicine. 10th ed. Saunders: Elsevier; 2015. p. 782.
196. Hanaoka M, Hisano M, Watanabe N, et al. Changes in the prevalence of the measles, rubella, varicella-zoster, and mumps virus antibody titers in Japanese pregnant women. Vaccine. 2013;31:2343–7.

197. Royal College of Obstetricians and Gynaecologists (RCOG). Chickenpox in pregnancy (Green-top guideline no. 13). London: Royal College of Obstetricians and Gynaecologists; 2015. https://www.rcog.org.uk/en/guidelines-research-services/guidelines/gtg13/.

198. Kotchmar Jr GS, Grose C, Brunell PA. Complete spectrum of the varicella congenital defects syndrome in 5-year-old child. Pediatr Infect Dis. 1984;3:142–5.

199. Lamont RF, Sobel JD, Carrington D, et al. Varicella-zoster virus (chickenpox) infection in pregnancy. BJOG. 2011;118:1155–62.

200. Mofenson LM. Advances in the prevention of vertical transmission of human immunodeficiency virus. Semin Pediatr Infect Dis. 2003;14:295–308.

201. Fowler MG, Lampe MA, Jamieson DJ, et al. Reducing the risk of mother-to-child immunodeficiency virus transmission: past successes, current progress and challenges, and future directions. Am J Obstet Gynaecol. 2007;197(Suppl 3):S3–9.

202. Newell ML. Current issues in the prevention of mother-to-child transmission of HIV-1 infection. Trans R Soc Trop Med Hyg. 2006;100:1–5.

203. Patel K, Hernàn MA, Williams PL, et al. Long-term effectiveness of highly active antiretroviral therapy on the survival of children and adolescents with HIV infection: a 10-year follow-up study. Clin Infect Dis. 2008;46:507–15.

204. Sripipatana T, Spensley A, Miller A, et al. Site-specific interventions to improve prevention of mother-to-child transmission of human immunodeficiency virus programs in less developed countries. Am J Obstet Gynaecol. 2007;197(Suppl 3):S107–12.

205. Shetty AK, Maldonado YA. Human immunodeficiency virus/acquired immunodeficiency syndrome in the infant. In: Remington JS, Klein JO, editors. Infectious diseases of the fetus and newborn infant. 7th ed. Philadelphia: Saunders; 2011. p. 622–60.

206. Thorne C, Newell ML. HIV. Semin Fetal Neonatal Med. 2007;12:174–81.

207. The European Collaborative Study. Vertical transmission of HIV-1: maternal immune status and obstetric factors. AIDS. 1996;10:1675–81.

208. The Working Group on Mother-to-Child Transmission of HIV. Rates of mother-to-child transmission of HIV-1 in Africa, America, and Europe: results from 13 perinatal studies. J Acquir Immune Defic Syndr Hum Retrovirol. 1995;8:506–10.

209. Garcia PM, Kalish LA, Pitt J, et al. Maternal levels of plasma human immunodeficiency virus type 1 RNA and the risk of perinatal transmission. Women and Infants Transmission Study Group. N Engl J Med. 1999;341:394–402.

210. Landesman SH, Kalish LA, Burns DN, et al. Obstetrical factors and the transmission of human immunodeficiency virus type 1 from mother-to-child. N Engl J Med. 1996;334:1617–23.

211. Tovo PA, de Martino M, Gabiano C, et al. Mode of delivery and gestational age influence perinatal HIV-1 transmission. J Acquir Immune Defic Syndr Hum Retrovirol. 1996;11:88–94.

212. Marion RW, Wiznia AA, Hutcheon G, et al. Human T-cell lymphotropic virus type III (HTLV-III) embryopathy: a new dysmorphic syndrome associated with intrauterine HTLV-III infection. Am J Dis Child. 1989;140:638–40.

213. Qazi QH, Sheikh TM, Fikrig S, et al. Lack of evidence for craniofacial dysmorphism in perinatal human immunodeficiency virus infection. J Pediatr. 1998;112:7–11.

214. Pollack H, Glasberg H, Lee E, et al. Impaired early growth of infants perinatally infected with human immunodeficiency virus: correlation with viral load. J Pediatr. 1997;130:915–22.

215. Kline MW. Vertically acquired human immunodeficiency virus infection. Semin Pediatr Infect Dis. 1999;10:147–53.

216. Di John D, Krasinski K, Lawrence R, et al. Very late onset of group B streptococcal disease in infants infected with the human immunodeficiency virus. Pediatr Infect Dis J. 1990;9:925–8.

217. Dumois JA. Potential problems with the diagnosis and treatment of syphilis in HIV-infected pregnant women. Pediatr. AIDS HIV Infect Fetus Adolesc. 1992;3:22–4.

218. American College of Obstetricians and Gynecologists (ACOG). The importance of preconception care in the continuum of women's health care. 2005 Committee Opinion number 313. Obstet Gynecol. 2005;106(3):665–6.

219. World Health Organization (WHO). Hormonal contraceptive methods for women at high risk of HIV and living with HIV: 2014 guidance statement. 2014. http://apps.who.int/iris/bitstream/10665/128537/1/WHO_RHR_14.24_eng.pdf?ua=1. Accessed 9 Sept 2016.

220. Loutfy MR, Blitz S, Zhang Y, et al. Self-reported preconception care of HIV-positive women of reproductive potential: a retrospective study. J Int Assoc Provid AIDS Care. 2014;13(5): 424–33.

221. Tubiana R, Le Chenadec J, Rouzioux C, et al. Factors associated with mother-to-child transmission of HIV-1 despite a maternal viral load <500 copies/ml at delivery: a case-control study nested in the French perinatal cohort (EPF-ANRS CO1). Clin Infect Dis. 2010;50(4): 585–96.

222. European Collaborative S. Mother-to-child transmission of HIV infection in the era of highly active antiretroviral therapy. Clin Infect Dis. 2005;40(3):458–65.

223. Panel on Treatment of HIV-Infected Pregnant Women and Prevention of Perinatal Transmission. Recommendations for use of antiretroviral drugs in pregnant HIV-1-infected women for maternal health and interventions to reduce perinatal HIV transmission in the United States. Updated 2015. 2015. https://aidsinfo.nih.gov/contentfiles/perinatalgl.pdf. Accessed 10 Sept 2016.

224. European Mode of Delivery Collaboration. Elective caesarean-section versus vaginal delivery in prevention of vertical HIV-1 transmission: a randomised clinical trial. Lancet. 1999;353(9158):1035.

225. World Health Organization (WHO). Consolidated guidelines on the use of antiretroviral drugs for treating and preventing HIV infection. 2nd ed. 2016. http://www.who.int/hiv/pub/arv/arv-2016/en/. Accessed 5 Sept 2016.

Influenza

2

Cheryl Cohen and Gary Reubenson

Abstract

Influenza is one of the commonest infections in human populations, and causing substantial morbidity and mortality globally. The influenza virus is divided into different types and subtypes, three of which are currently circulating widely in humans: influenza A(H3N2) and influenza B. The virus undergoes constant evolution, leading to annual seasonal winter epidemics in temperate countries and necessitating annual updates to the vaccine. Rarely, completely new influenza viruses can emerge in human populations, giving rise to influenza pandemics. Children aged <5 years (especially those <2 years) and those with underlying illness such as cardiac, respiratory and severe neurologic disease have an increased risk of severe outcomes associated with influenza. Pregnant women have an increased risk of severe influenza. Complications may involve the respiratory tract (e.g. otitis media or pneumonia) or, less commonly, other organ systems (e.g. encephalitis or myocarditis). Specific antiviral treatment should be offered as soon as possible for hospitalized children with presumed or confirmed influenza and for influenza of any severity for children at high risk of severe complications of influenza without waiting

C. Cohen, Ph.D. (✉)
Centre for Respiratory Diseases and Meningitis, National Institute for Communicable Diseases of the National Health Laboratory Service, Johannesburg, South Africa

School of Public Health, Faculty of Health Sciences, University of the Witwatersrand, Johannesburg, South Africa
e-mail: cherylc@nicd.ac.za

G. Reubenson, M.D.
Department of Paediatrics and Child Health, Faculty of Health Sciences, Empilweni Service & Research Unit, Rahima Moosa Mother and Child Hospital, University of the Witwatersrand, Johannesburg, Gauteng, South Africa

© Springer International Publishing AG 2017
R.J. Green (ed.), *Viral Infections in Children, Volume I*,
DOI 10.1007/978-3-319-54033-7_2

for laboratory confirmation. Antiviral treatment is usually not warranted for uncomplicated influenza as this is usually self-limiting. Annual influenza vaccination should be offered to all individuals at increased risk for complications of influenza. Vaccine cannot be given to children aged <6 months but maternal influenza immunization during pregnancy is recommended and can confer protection to the young infant.

2.1   Introduction

Influenza is one of the commonest infections in human populations, infecting a significant percentage of the population each year and causing substantial morbidity and mortality globally. While seasonal influenza is an important cause of morbidity and mortality, novel influenza virus strains can emerge with the potential to cause global pandemics. Children are a common source of inluenza transmission in the community and form an important risk group for severe influenza illness (particularly infants and children with underlying chronic illnesses).

2.2   Virological Data

Influenza viruses are enveloped viruses from the family *Orthomyxoviridae* [1]. Influenza has a negative sense RNA genome divided into eight segments. The segmented genome allows for exchange of genes between influenza viruses of the same type through genetic reassortment. Two types of influenza viruses, influenza A and B, cause epidemic disease in humans. The influenza A viruses are divided into subtypes and influenza B viruses into lineages based on their antigenic structure. The influenza A subtypes are differentiated based on characteristics of the haemagglutinin (HA) and neuraminidase (NA) surface antigens. Haemaglutinin is responsible for virus attachment during the early stages of infection and is the main antigen against which the host immune response is directed. Neuraminidase facilitates the release of mature virus from the cell surface.

Currently there are two influenza A subtypes (influenza A(H3N2) and influenza A(H1N1)pdm09 and two influenza B lineages (Yamagata and Victoria) co-circulating globally in human populations. Influenza A viruses of at least 17 HA and 9 NA subtypes have been isolated from animals such as birds, pigs, horses and dogs. Because many different HA and NA subtypes can circulate in animals, animals such as birds and pigs may be the reservoir of emerging influenza virus subtypes which can infect humans [1]. Several of these subtypes such as influenza A(H7N1) or influenza A(H5N2) can cause severe illness and even death in individuals in close contact with animals but are not able to be efficiently transmitted from person-to-person [2].

Antigenic drift is the emergence of new influenza virus antigenic variants as a result of point mutations and recombination which occurs during viral replication [1]. This frequent emergence of antigenic variants contributes to seasonal influenza epidemics and leads to the requirement for annual assessment of the need to update

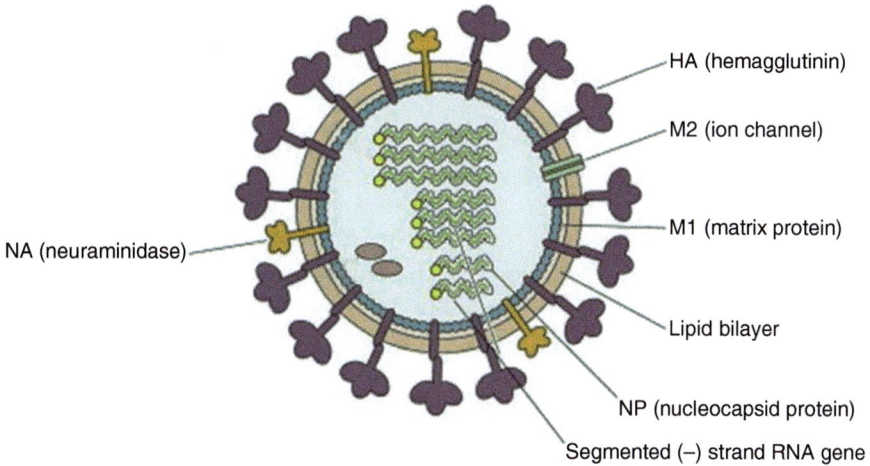

Fig. 2.1 The structure of influenza virus. Credit to virology blog: about viruses and viral disease by Vincent Racaniello http://www.virology.ws/2009/04/30/structure-of-influenza-virus/

the viruses included in the influenza vaccine. Antigenic shift is a term for larger genetic changes which occur infrequently in influenza A viruses. New, or substantially different, influenza A virus subtypes which emerge in humans, have the potential to cause pandemics if they are efficiently transmitted between humans in the presence of little or no pre-existing population immunity [2] (Fig. 2.1).

2.3 Epidemiology Including Pandemics

2.3.1 Burden of Disease

It is estimated that 5–20% of the population become infected with influenza each year and about 20% of these develop symptomatic illness. Rates of influenza infection are highest in children aged 5–15 years [3]. Annually in children aged <5 years there are approximately 90 million new cases of influenza, 20 million cases of influenza-associated acute lower respiratory tract infection (ALRI) (13% of all cases of paediatric ALRI) and one million cases of influenza-associated severe ALRI (7% of all severe ALRI) globally [4]. Between 28,000 and 111,500 influenza-associated deaths in children <5 years are estimated to occur each year, with 99% of these occurring in developing countries. Influenza is associated with approximately 10% of respiratory hospitalizations in children <18 years worldwide ranging from 5% in children aged <6 months to 16% in children aged 5–17 years [5]. Influenza-associated hospitalization rates are more than three times higher in developing than industrialised countries. The incidence and mortality associated with influenza can vary substantially from year-to-year as a result of different circulating types and subtypes with differing propensity to cause severe illness. Years in which influenza A(H3N2) predominates may typically be associated with increased risk of severe disease [6].

People of all ages may develop symptomatic influenza infection but the highest rates of influenza-positive influenza-like illness (ILI) are seen in children aged 2–17 years [7]. School-age children are an important source of infection in the community and influenza outbreaks can occur in schools during the influenza season [8]. During the influenza season, influenza is an important cause of school absenteeism. Illness in children can cause a substantial economic burden as a result of caregiver absenteeism from work to care for ill children as well as outpatient visits in children and can lead to additional antibiotic courses being prescribed. Hospitalizations and mortality during the influenza season can be substantial. In severe influenza seasons, the large number of medical care visits as a result of influenza can overwhelm health systems.

2.3.2 Groups at Risk for Severe Disease

The highest rates of influenza-associated hospitalizations and deaths are typically seen in individuals aged ≥65 years, <5 years and those with underlying medical conditions that confer an increased risk for severe influenza [9]. Children aged <2 years and, to a lesser extent, those aged 2–5 years have increased rates of influenza-associated hospitalization and mortality compared to older children. Children with underlying illnesses, particularly cardiac, respiratory and severe neurologic disease have an increased risk of severe outcomes associated with influenza. A study from South Africa, found that amongst children aged <5 years, malnutrition, prematurity and HIV infection were associated with increased odds of influenza-associated hospitalization [10]. HIV-infected children have an approximately two times elevated risk of influenza hospitalization and are more likely to die of influenza once hospitalized compared to HIV-uninfected children [11, 12].

Pregnant women have an increased risk of severe influenza. Some studies suggest that influenza in pregnancy may be associated with adverse outcomes in infants born to these women (such as low birth weight, pre-term birth and stillbirth), but others have disputed this [13].

2.3.3 Seasonality

In temperate climates influenza typically causes annual seasonal epidemics in the winter months, between April and September in the Southern Hemisphere and between October and April in the Northern Hemisphere [9, 14]. In more tropical climates influenza commonly circulates year-round with two or more peaks which may coincide with climatic events such as the rainy season [15]. This may present challenges for decision-making around the best time to vaccinate and which vaccine formulation (the Northern or Southern Hemisphere) should be used (see section on vaccines) [16]. The start, peak, size and duration of the influenza season may vary substantially from year-to-year. Seasonal influenza can give rise to outbreaks in closed settings such as schools, these can occur at any time of year but are more common during the influenza season [9].

2.3.4   Pandemic Influenza

Influenza pandemics are caused by the emergence and spread in human populations of a new influenza A virus with either a new or substantially altered HA or NA combination against which there is little or no immunity in humans, which is easily transmitted between humans and causes clinical illness in humans [1]. The emergence of a pandemic influenza strain is unpredictable and can occur through two mechanisms. Firstly, a host could be simultaneously infected with two different influenza virus subtypes which could allow for exchange of genetic material or reassortment and the emergence of a new subtype. For example this could occur if a pig were infected by both a human and avian origin influenza subtype simultaneously with genetic exchange leading to the emergence of a virus adapted to spread in humans, but with HA and/or NA not currently circulating in humans. The second way that novel subtypes can emerge is if avian or other animal adapted subtypes are directly transmitted to humans and then undergo adaptation to allow transmission between humans. Currently some avian influenza virus subtypes such as influenza A(H5N2) can be transmitted to humans, usually following close contact with poultry, and cause severe infections. However, these viruses are not adapted for efficient transmission from person to person and therefore have not given rise to a new pandemic strain [2]. Global surveillance for new influenza virus strains is essential for early identification of novel strains to allow a global public health response.

The 1918 pandemic of influenza A(H1N1) is widely acknowledged as the most severe in recent times with an estimated >20 million deaths worldwide. Other recent pandemics (1957, Asian flu H2N2 and 1968 Hong Kong flu H3N2) have been associated with a lower death toll [17]. A characteristic of pandemic influenza strains is the shift in the age distribution of deaths from predominantly affecting the extremes of age (young infants and the elderly) to mortality in young adults aged 20–40 years [17]. Although influenza pandemics can cause substantial mortality, the annual cumulative deaths each year, associated with seasonal mortality, far outweigh this burden.

In 2009, a novel influenza A virus, influenza A(H1N1)pdm09 emerged in the human population and caused a global pandemic. This virus, was antigenically distinct from the H1N1 virus which had been circulating in human populations from 1997 to early 2009 and was thought to have entered the human population from pigs (hence the colloquial name "swine flu"). The overall mortality burden of this strain was estimated at between 123,000 and 203,000 deaths globally, similar to the annual mortality burden from seasonal influenza, although this strain did exhibit the characteristic pandemic age shift, disproportionately affecting individuals aged 20–40 years [18]. Subsequently, influenza A(H1N1)pdm09 has been circulating in human populations and immunity in the population has built up. Influenza A(H1N1)pdm09 has become the predominant H1N1 seasonal train, replacing those that previously circulated and behaves like any other seasonal influenza virus [9].

2.4   Transmission and Pathogenesis of Disease

Influenza is predominantly spread person-to-person by large droplets and through direct contact with respiratory secretions [19]. The contribution of airborne transmission is unclear. The incubation period for influenza typically ranges from 1 to 4 days (median 2 days) [20]. Influenza virus is typically shed from the nasopharynx for up to 5 days after illness but viral shedding may be longer in severely ill individuals, young children and immunocompromised individuals. The reproductive number for influenza is between 1 and 2 and the serial interval usually estimated at 2–3 days.

An individual's susceptibility to infection and disease will depend on host characteristics including preexisting cellular or humoral immunity to influenza [20]. Young children may have no pre-existing immunity to influenza, but older children and adults have often been exposed to circulating influenza several times before and may also have pre existing immunity from vaccination. Natural immunity is not fully protective, largely because of the variability of influenza HA and NA.

Influenza virus replication predominantly occurs in the respiratory tract columnar epithelial cells, with infection leading to loss of cilia and cell death [20]. Damage to the respiratory tract as well as immunologic changes can lead to increased susceptibility to bacterial superinfection. Viremia with influenza is relatively uncommon although constitutional symptoms are a prominent feature of clinical disease.

2.5   Clinical Manifestations

In most children influenza infection results in acute self-limiting upper respiratory tract (URT) symptoms, however, systemic manifestations are not uncommon [21]. Factors that influence clinical presentation include: age of the child, previous influenza exposure, vaccination status, underlying disease states or co-morbidities, as well as viral factors.

Children are considered important influenza "vectors" and are often responsible for introducing the virus into their homes and broader social settings [22].

Influenza classically presents with the sudden onset of systemic (fever, myalgia, headache, and malaise) and URT symptoms (sore throat, cough, rhinitis). Since many patients do not have all these typical symptoms, accurate clinical diagnosis is challenging particularly in the younger pre-verbal child and outside of the influenza season [23].

A large study evaluating the clinical presentation of influenza in children found that almost all (95%) had fever; cough (77%) and rhinitis (78%) were also very common, but much lower proportions experienced headache (26%) or myalgia (7%) [24]. Younger children have not yet been exposed to influenza very often and so have yet to acquire immunity to a substantial repertoire of circulating seasonal influenza strains. They, therefore, are more likely to develop severe or complicated disease [25]. Further, they are less likely to manifest with classic symptoms, experience higher fevers (not uncommonly associated with febrile convulsions), less prominent URT involvement and more gastro-intestinal symptoms (vomiting, diarrhea, abdominal pain, loss of appetite).

Examination may be completely normal in some children, others may manifest with tachypnea, conjunctival injection, nasal inflammation and discharge, or cervical lymphadenopathy. Oropharyngeal findings are often limited, even in those children complaining of a sore throat [24].

Symptoms of uncomplicated influenza usually start improving within a few days, but symptoms lasting more than a week are not uncommon. Cough, in particular, may persist for a number of weeks, but steady improvement can be expected [26].

The differential diagnosis of influenza largely depends on the presenting symptoms and clinical findings, but includes other respiratory viruses (rhinovirus, coronavirus, respiratory syncitial virus, human metapneumovirus, adenovirus, parainfluenza) and some bacterial URT infections (*Streptococcus pyogenes*, Mycoplasma). The clinical manifestations of these conditions are very similar, regardless of the implicated pathogen [27]. All influenza strains may result in severe illness and knowing which infection a particular child has, is not helpful in predicting their disease course.

Complications may involve the respiratory tract (e.g. otitis media, pneumonia) or, less commonly, other organ systems (e.g. encephalitis, myocarditis). Otitis media may occur in as many as 50% of cases; this may be related to the influenza virus itself or secondary infection with bacteria or other viruses [24, 28]. Symptoms of acute otitis media generally present a few days after onset of influenza symptoms.

Lower respiratory tract complications may include the following [21, 29]:

- Laryngo-tracheo-bronchitis ("croup")
- Bronchiolitis
- Pneumonia—especially in children <2 years of age, often mild but may be severe, rapidly-progressive and occasionally fatal, particularly if associated with secondary bacterial infection (usually *Streptococcus pneumoniae* or *Staphylococcus aureus*). A variety of radiographic appearances have been described
- Acute exacerbation of asthma—this is the most common respiratory tract complication of influenza.

Central nervous system involvement can include the following [30–32]:

- Aseptic meningitis
- Acute cerebellitis
- Transverse myelitis
- Guillain-Barré syndrome
- Febrile seizure
- Necrotizing encephalitis
- Postinfectious encephalitis (also referred to as acute disseminated encephalomyelitis).

Neurologic complications appear to be more common in younger children and in those with underlying neurologic and neuromuscular disease. Following the rapid decline in aspirin use over the last few decades, influenza-associated Reye syndrome is now rare.

Mild transient myositis is common with influenza infection; it is more likely with influenza B and is associated with moderate elevations in creatine kinase levels [24]. Acute myositis is an important, severe, but rare, complication of influenza infection [33]. It presents with extreme muscle tenderness, often involving the calf muscles, extreme elevations in creatine kinase as well as significant myoglobinuria.

2.5.1 Diagnostic Testing

During influenza season, influenza should be considered in all children presenting with suggestive clinical features—this includes those already admitted to hospital, as nosocomial transmission of influenza is well described. Influenza should still be considered outside of the influenza season, particularly in travelers and children residing in tropical and sub-tropical climates where year-round influenza transmission occurs.

Accurate clinical diagnosis of influenza is challenging, particularly in younger children. The lack of specific signs or symptoms results in patients receiving diagnoses of "influenza-like illness" or "viral upper respiratory tract illness" unless further diagnostic testing is undertaken. This degree of diagnostic uncertainty should be acknowledged, however, since most such cases are self-limiting and management is largely supportive, there is usually no need to obtain a precise microbiological diagnosis.

Currently available diagnostic tests for influenza include the following [23, 34]:

- **Rapid, point-of-care, antigen detection tests**
  A number of different tests are currently available, although they remain unavailable in many settings. They provide results within 30 min, and, when used appropriately, are helpful in confirming influenza infection. In general, they are insufficiently sensitive to reliably exclude the disease. Further, their performance will depend on which antigens are expressed by currently circulating strains. When influenza activity is low, positive results are likely to be false-positives, however, their positive predictive value improves as influenza activity increases. Conversely, during periods of high influenza activity, false-negatives are more likely and may warrant additional testing in some patients.

  Since diagnostic confirmation seldom affects management of such children, the use of these tests is generally not recommended in low-resource settings and should be used judiciously in better-resourced areas. When used, it should be clear in the clinician's mind as to how the result is going to alter treatment: positive results can potentially reduce antibiotic usage and allow for early use of antivirals in those at high risk of complications or severe disease. However, it is important to recognize that the identification of influenza does not exclude the presence of bacterial co-infections.

- **Polymerase Chain Reaction (PCR) tests**
  These tests are currently considered to be the most reliable for the diagnosis of influenza in children. Amongst the available options, they are the most sensitive and specific. They can be performed on most respiratory samples, most commonly

nasopharyngeal aspirates or swabs. They are also able to differentiate influenza A and B, as well as subtypes of influenza A.

More recently, point-of-care PCR assays have becomes available in some developed world settings. They are performed on nasal swabs and can supply a reliable result in as little as 15 min. While, not currently available in most developing world settings, if more affordable they could become an important new influenza diagnostic.

Since the live attenuated influenza vaccine contains influenza genetic material, recent receipt of this vaccine will also result in a positive PCR test.

- **Immunofluorescence tests**
These are also performed on nasal or nasopharyngeal swabs and allow for the direct or indirect detection of influenza antigens. Influenza A and B can be differentiated, however, the sensitivity of these tests is moderate and, particularly during periods of high transmission, negative tests may need to be repeated using a more sensitive methodology (PCR or culture).
- **Viral culture**
Viral culture takes at least 48–72 hours and so has limited utility in the routine diagnosis of influenza. However, they are helpful as part of surveillance activities and isolates can be used to inform annual vaccine planning.
- **Serological assays**
A variety (hemagglutination-inhibition, enzyme-linked immunosorbent assay (ELISA), and complement fixation assays) of serological assays can be performed but are of limited diagnostic value as they require acute and convalescent sampling. A fourfold increase in titre allows for a retrospective influenza diagnosis to be made. Their primary role is as a research tool.

2.6 Therapeutic Options

Antivirals are available for the specific treatment of influenza. Two classes of anti-influenza drugs are available: neuraminidase inhibitors (e.g. oseltamivir, zanamivir) and M2 inhibitors (e.g. amantadine, rimantadine). Both classes are inactive against all other respiratory viruses. Resistance can emerge in circulating influenza virus strains and for this reason antiviral resistance should be constantly monitored through surveillance and recent guidance consulted for the latest antiviral resistance profiles.

- **Neuraminidase inhibitors** [35]
Neuraminidase inhibitors prevent the release of new virions from influenza infected cells and are active against both influenza A and B.

Oseltamivir is the most widely available member of this drug class and can be used for children and adults. It is dosed orally and is approved for both treatment and prevention of influenza. It is available both in capsule form and as a powder for suspension, although the suspension has a relatively short shelf life and is often substantially more expensive than the capsule. As a result, the capsule is the most widely used formulation. When the oral suspension is unavailable the capsule can be opened and diluted with sweetened liquids to provide the appropriate dose.

Zanamivir is predominantly available as an inhaled formulation, although intravenous zanamivir is available for investigational use, particularly for severely ill patients or those with suspected or confirmed oseltamivir-resistant virus. The inhaled preparation is contra-indicated in children with a history of wheezing or other chronic respiratory condition. Its use is not recommended for children younger than 5 years of age.

Peramivir is an intravenous neuraminidase inhibitor that is not approved for use in children and is not widely available.

Neuraminidase inhibitors are generally well tolerated. Common side effects include nausea, vomiting and rash, as well as bronchospasm with zanamivir. Neuropsychiatric symptoms have been linked to oseltamivir use, particularly in Japan, however, more recent evidence suggests no such causative link [36, 37]. Severe adverse reactions have been reported but are considered rare [38].

The mainstay of influenza prevention remains immunization, but in high risk situations, amongst partially or unimmunized children, chemoprophylaxis can be considered. Both pre- and post-exposure chemoprophylaxis have been advocated, however, concerns regarding induction of oseltamivir-resistance have tempered enthusiasm for this approach and their prophylactic use is increasingly discouraged.

Antiviral treatment should be offered as soon as possible for hospitalized children with presumed or confirmed influenza and for influenza of any severity for children at high risk of severe complications of influenza (Table 2.1) [39]. Timely oseltamivir treatment can reduce the duration of fever and symptoms. There are no prospective randomized controlled trials of treatment efficacy of neuraminidase inhibitors in hospitalized children and for severe outcomes but observational data suggest that treatment does reduce the risk of hospitalization and death, although there is some controversy in the literature (Table 2.2) [40]. Treatment should be initiated as early as possible. The benefit of any treatment is maximal early in the course of the disease and should be started within 48–72 hours of symptom onset. Influenza diagnosis may not have been confirmed within this timeframe and so treatment will often be initiated empirically prior to microbiological confirmation. Some evidence suggests benefit to those with very severe illness even when treatment is initiated later than 72 hours into the disease course. Therefore, in

Table 2.1 Children at high risk of severe influenza in whom influenza antiviral treatment is recommended by the Centers for Disease Control and Prevention (CDC) and American Academy of Pediatrics (AAP) current guidance [9, 39]

1. Children aged <2 years
2. Persons with chronic pulmonary (including asthma), cardiovascular (except hypertension alone), renal, hepatic, hematologic (including sickle cell disease), or metabolic disorders (including diabetes mellitus) or neurologic and neurodevelopment conditions (including disorders of the brain, spinal cord, peripheral nerve, and muscle such as cerebral palsy, epilepsy [seizure disorders], stroke, intellectual disability, moderate to severe developmental delay, muscular dystrophy, or spinal cord injury)
3. Persons with immunosuppression, including that caused by medications or by HIV infection
4. Persons who are receiving long-term aspirin therapy
5. American Indian/Alaska Native persons
6. Residents of chronic care facilities

Table 2.2 Currently recommended doses for neuraminidase inhibitors (modified from [39])

Medication	Treatment (5 days)	Chemoprophylaxis (10 days)
Oseltamivir		
Children aged ≥12 months		
Body weight		
<15 kg (≤33 lb)	30 mg twice daily	30 mg once daily
>15–23 kg (33–51 lb)	45 mg twice daily	45 mg once daily
>23–40 kg (>51–88 lb)	60 mg twice daily	60 mg once daily
>40 kg (>88 lb)	75 mg twice daily	75 mg once daily
Infants aged 9–11 months	3.5 mg/kg per dose twice daily	3.5 mg/kg per dose once daily
Term infants aged 0–8 months	3 mg/kg per dose twice daily	3 mg/kg per dose once daily for infants 3–8 months; not recommended for infants <3 months old unless situation is judged critical, because of limited safety and efficacy data in this age group
Preterm infants	1 mg/kg per dose, orally, twice daily, for those <38 weeks' postmenstrual age	
	1.5 mg/kg per dose, orally, twice daily, for those 38 through 40 weeks' postmenstrual age	
	3.0 mg/kg per dose, orally, twice daily, for those >40 weeks' postmenstrual age	
	For extremely preterm infants (<28 weeks), consult a pediatric infectious diseases physician	
Zanamivir		
Children (≥7 years for treatment, ≥5 years for chemoprophylaxis)	10 mg (two 5-mg inhalations) twice daily	10 mg (two 5-mg inhalations) once daily

patients with severe or complicated disease, treatment should be initiated even if >48 hours after illness onset. Children meeting the clinical criteria for treatment should be treated irrespective of whether they have been vaccinated.

Since influenza is generally a mild, self-limiting disease in previously well children, most such children will not require antiviral treatment even when they present soon after symptom onset. Treatment of such children increases the risk of adverse events, potentially increases the risk of resistance developing and may deplete medicine supply for those in greater need.

• **Adamantanes** [9, 39]

These agents target the influenza A M2 protein, which is essential for efficient viral replication. They have no activity against influenza B and are not active against currently circulating influenza A strains. As such, there use is not currently recommended for the treatment or prevention of influenza. It has been suggested that they may have a role, in combination with oseltamivir, for the treatment of oseltamivir-resistant influenza A.

2.7   Vaccines and Guidelines for Vaccination

2.7.1   Process of Annual Vaccine Selection

Because of the changing nature of influenza viruses, the World Health Organization (WHO) monitors the epidemiology of influenza viruses throughout the world through the Global Influenza Surveillance and Response System (GISRS). Separate recommendations are made for the Southern Hemisphere and Northern Hemisphere vaccine strains each year. Each year, towards the end of the influenza season, recommendations about strains to be included in the vaccine for each Hemisphere for the following influenza season are made.

2.7.2   Groups Recommended to Receive Annual Influenza Vaccination

Influenza vaccination can be given to any person who wishes to reduce the risk of becoming ill during the influenza season. Some countries such as the United States of America (USA) and United Kingdom (UK) recommend influenza vaccination for all children, or all individuals. In addition, special effort should be made to vaccinate children at risk of severe influenza listed in Table 2.3. Individuals such as healthcare personnel and childcare providers (especially those in contact with infants aged <6 months and children with underlying risk conditions) should be vaccinated to reduce the risk of transmission to high risk children. Lastly, pregnant women are recommended to receive influenza vaccination, to reduce the risk of severe illness in the mother, to provide direct protection to the young infant through trans-placental transfer of maternal antibodies and to reduce the risk of transmission of influenza from the mother to the young infant [41].

Table 2.3 Groups recommended for influenza vaccination of particular relevance in pediatrics (adapted from [39])

A. Children and adolescents at increased risk of severe influenza
1. All children and adolescents, including infants born preterm, who are 6 months and older (based on chronologic age) with conditions that increase the risk of complications from influenza including:
a. Asthma or other chronic pulmonary diseases, including cystic fibrosis
b. Hemodynamically significant cardiac disease
c. Immunosuppressive disorders or therapy

Table 2.3 (continued)

d. HIV infection
e. Sickle cell anemia and other hemoglobinopathies
f. Diseases that necessitate long-term aspirin therapy, including juvenile idiopathic arthritis or Kawasaki disease
g. Chronic renal dysfunction
h. Chronic metabolic disease, including diabetes mellitus
i. Any condition that can compromise respiratory function or handling of secretions or can increase the risk of aspiration, such as neurodevelopmental disorders, spinal cord injuries, seizure disorders, or neuromuscular abnormalities
j. Morbid obesity
k. Pregnancy
2. Children aged 6 months–59 months as this group (particularly those aged <24 months) has an increased risk of influenza-associated hospitalization and mortality
B. Individuals who come into contact with children at risk of severe influenza
1. All household contacts and out-of-home care providers of the following: children with high-risk conditions; and children younger than 5 years, especially infants younger than 6 months
2. All health care personnel (HCP)
3. All child care providers and staff; and
4. All women who are pregnant, are considering pregnancy, are in the postpartum period, or are breastfeeding during the influenza season

2.7.3 Available Influenza Vaccines (Table 2.4)

2.7.3.1 Trivalent Inactivated Vaccine (IIV3) and Quadrivalent Inactivated Vaccine (IIV4)

IIV3 has been available for many years and includes inactivated components of two influenza A (one each of influenza A(H1N1)pdm and influenza A(H3N2)) and one influenza B strains. Since the 1980s, two antigenically distinct influenza B lineages (Victoria and Yamagata) have been circulating globally. This is a limitation of IIV3, as protection may be reduced when the circulating influenza B strain is of the lineage which is not included in IIV3. The IIV4 includes an additional strain of the other influenza B lineage not included in TIV to make a total of four strains and thus potentially offers additional benefit of protection against both circulating influenza B lineages.

Inactivated influenza vaccines (IIV3 and IIV4) contain no live virus. Standard-dose IIV should contain 15 μg of each haemagglutinin antigen in each 0.5 mL dose. IIV3 and IIV4 are available in formulations for both intramuscular (IM) and intradermal (ID) use but the ID formulation is only licensed for use in individuals aged 18 years and older.

Two formulations of IIV3 manufactured using technologies that do not include eggs have become available in recent years, but neither is licensed for individuals aged <18 years. These are cell-culture based inactivated influenza vaccine and recombinant influenza vaccine.

Table 2.4 Types of influenza vaccines available for use in children (note this list is not comprehensive and other formulations may be available in some settings)

Type	Trade name	Manufacturer	Presentation	Age indications	Route
Trivalent IIV standard dose	Vaxigrip®	Sanofi Pasteur	0.5 mL liquid in a single-dose prefilled syringe	≥6 months	IM
	Influvac®	Abbott	0.5mL single dose prefilled syringe	≥6 months	IM
	Vaxigrip®	Sanofi Pasteur	5.0 mL multi-dose vial	≥6 months	IM
	Fluvac®	bioCSL	0.5 mL single dose prefilled syringe	≥9 years	IM
			5.0 mL multidose vial	≥9 years	IM
	Fluvirin®	Novartis vaccines and diagnostics	0.5 mL single-dose prefilled syringe	≥4 years	IM
			5.0 mL multidose vial	≥4 years	IM
Quadrivalent IIV standard dose	Fluzone®	Sanofi Pasteur	0.25 mL single dose prifilled syringe	6–35 months	IM
			0.5 mL single-dose prefilled syringe/0.5 single dose vial	≥36 months	IM
			5.0 mL multidose vial	≥6 months	IM
	Fluarix®	GlaxoSmithKline (GSK)	0.5 mL single dose prefilled syringe	≥3 years	IM
	Flulaval®	ID Biomedical Corp. of Quebec (distributed by GSK)	5.0 mL multidose vial	≥ 3 years	IM
Live attenuated influenza vaccine	FluMist Quadrivalent®	MedImmune	0.2 mL single-dose prefilled intranasal sprayer	2–49 years	IN

IN intranasal, *IM* intramuscular, *IIV* inactivated influenza vaccine

2.7.3.2 Live-Attenuated Influenza Vaccine (LAIV)

LAIV is a live attenuated influenza vaccine which is administered intranasally and licensed for use in individuals aged 2–49 years of age. Since 2013 the LAIV has only been available in a quadrivalent formulation.

2.7.3.3 Adjuvanted Influenza Vaccine

Adjuvanted formulations of influenza vaccine are licensed for use in individuals aged ≥65 years in the USA but not currently in children [39]. They have been shown

Table 2.5 Recommended dosage of influenza vaccine for patients of different age groups

Age group	Dose	Number of doses
Adults and children 9 years of age and older	Adult dose (0.5 mL) IMI	Single dose
Children 3 years–8 years	Adult dose (0.5 mL) IMI	1 or 2 doses[a]
Children 6 months–2 years	0.25 mL (half an adult dose) IMI	1 or 2 doses[a]

Note: influenza vaccine is not recommended for infants <6 months of age. *IMI* intramuscular injection
[a]For individuals who have not previously received a total of ≥2 doses, or when vaccine status is unknown, two doses should be administered ≥1 month apart

to have a higher efficacy then IIV in a randomized controlled trial in children [42]. Adjuvants have several potential advantages over more traditional vaccine formulations including increased immunogenicity, potentially reducing the amount of antigen required. They elicit a more robust immune response and could potentially reduce the number of doses needed in children.

2.7.4 Influenza Vaccine Dosage and Administration

Children aged 6 months through 8 years should receive two influenza doses administered ≥4 weeks apart the first time influenza vaccine is administered. For young children who require two doses of influenza vaccine, vaccination should not be delayed to ensure that both doses are given with the same product. Any licensed, effective influenza vaccine product may be used for each dose. It is important to document all doses of influenza vaccine administered in the child's medical records. In temperate countries, influenza vaccine should be administered as soon as possible after the influenza vaccine becomes available.

The recommended dosage of influenza vaccine for patients of different age groups is described in Table 2.5 [9].

2.7.5 Influenza Vaccine Effectiveness

Influenza vaccine effectiveness depends on characteristics of those being vaccinated (age and health), whether there is a good match between the circulating viruses and the viruses contained in the vaccine, and on influenza types and subtypes circulating each year. In general, influenza vaccines work best among children ≥2 years and healthy adults. Older people (≥65 years), children <2 years and severely immunocompromised individuals often have poorer immune responses to inactivated influenza vaccine (IIV) compared with healthy adults. However, even for these people influenza vaccine still provides some protection. Other products, e.g. high-dose influenza vaccine and adjuvanted vaccines, have been shown to be more effective in certain groups [43] but these vaccines may not be available in all settings and are not licensed for use in all age groups.

There have been a number of studies of IIV effectiveness in children aged 6–59 months. For seasonal IIV in young children, two doses of influenza vaccine provides better protection than one dose in the first season a child is vaccinated. Estimates of IIV efficacy in young children are limited and vary by season and study design. Efficacy is lower in children aged 6–23 months. Data are unclear as to the effectiveness in HIV-infected children aged <5 years [44].

A randomized controlled trial of LAIV3 in healthy children aged 15–71 found a vaccine effectiveness of 92% (95% CI 65–96%). Several other randomized controlled trials and observational studies have demonstrated high efficacy of LAIV3 against laboratory-confirmed influenza. Studies comparing efficacy of IIV and LAIV have generally found that LAIV has similar, or better efficacy, than IIV. Since 2013, LAIV4 has replaced LAIV3 as the available LAIV. Licensure was on the basis of immunogenicity studies. In 2016, conflicting findings on the effectiveness of LAIV emerged from different settings, with the USA withdrawing its recommendation for LAIV use in the 2016–2017 influenza season based on concerns of reduced effectiveness, particularly against the A(H1N1)pdm component of the vaccine [9]. Studies from Europe in the same seasons found moderate LAIV effectiveness, similar to that of IIV for the same year [45, 46] with recommendations for LAIV remaining unchanged in England and Europe.

Vaccinating individuals at risk of severe influenza may provide direct protection for these individuals. In addition, vaccinating individuals in close contact with people at risk for severe influenza may provide indirect protection through preventing transmission to high-risk individuals. Vaccinating children can protect children directly and the general population indirectly. This strategy is especially important for individuals in whom influenza vaccine is not indicated, such as children aged <6 months (who may be protected through maternal immunization) [47–49]. A randomized controlled trial conducted in South Africa has shown that when pregnant women receive the influenza vaccine, their risk of developing influenza is halved, as is the risk to their infants in the first 24 weeks of life [47]. The vaccine has been shown not only to be efficacious for prevention of influenza in both mothers and their infants, but also safe [47–49]. Vaccination of healthcare workers may decrease the risk of spreading influenza to their patients. Recent influenza vaccination does not preclude a diagnosis of influenza as the vaccine is not 100% effective.

Because of the large year-to-year variability in influenza vaccine effectiveness depending on the circulating influenza strains, many countries publish annual estimates of influenza vaccine effectiveness using a test-negative case-control study design. A systematic review of test-negative case-control studies found that the pooled VE was 33% (95% CI 26–39) for H3N2, 54% (46–61) for type B, 61% (57–65) for H1N1pdm09, suggesting reduced protection of available vaccines against influenza A(H3N2) in recent years [50].

2.7.6 Contraindications to Influenza Vaccination [35, 39]

The IIV is an inactivated vaccine, and has a well-established safety record. It is safe for use in pregnancy and in children ≥6 months of age. Minor illness, with or without fever, is not a contraindication to influenza vaccine administration. Clinicians should always consult the manufacturer's package insert for current contraindications and precautions for particular products.

Contraindications to the administration of IIV include:

- A history of severe (anaphylactic) hypersensitivity to any components of the vaccine including, egg protein, or after previous dose of any influenza vaccine. Anaphylaxis is rare and a careful history will distinguish between anaphylactic reactions and allergic reactions like rashes. Mild egg protein allergy is not a contraindication for influenza vaccine
- Infants <6 months of age.

Precautions to IIV administration include:

- Persons with moderate to severe illness with or without fever should preferably be immunized after symptoms have resolved
- Person who developed Guillian-Barrè syndrome within 6 weeks of receiving an influenza vaccine.

Contraindications to the administration of LAIV include the following:

- Children 2–4 years of age with a history of recurrent wheezing or a medically attended wheezing episode in the previous 12 months because of the potential for increased wheezing after immunization
- Children with the diagnosis of asthma
- Children with a history of egg allergy
- Children who have received other live virus vaccines within the past 4 weeks; however, other live virus vaccines can be given on the same day as LAIV
- Children who have known or suspected immunodeficiency disease or who are receiving immunosuppressive or immunomodulatory therapies
- Children who are receiving aspirin or other salicylates
- Any female who is pregnant or considering pregnancy
- Children with any condition that can compromise respiratory function or handling of secretions or can increase the risk for aspiration
- Children taking an influenza antiviral medication (oseltamivir or zanamivir), until 48 hours after stopping the influenza antiviral therapy
- Children with chronic underlying medical conditions that may predispose to complications after wild-type influenza infection.

As for all vaccines, influenza vaccine should be administered in a setting where there is the ability to respond to acute hypersensitivity reactions.

2.7.7   Influenza Vaccine Adverse Events

The most common adverse events following intramuscular IIV administration are pain and tenderness at the injection site. Fever can occur in 10–35% of children aged <2 years within 24 hours of vaccination, but is much less common in older children. In trials, when IIV are administered, 16–20% of those vaccinated experience local reactions in the arm, lasting for 1 or 2 days. Short-term reactions (mild fever, malaise and muscle pains) have been reported in a much smaller proportion in the first few hours following vaccination. Trials of the split and subunit vaccines show even fewer systemic reactions. There have been no strong temporal or causal associations of the current vaccines with more severe reactions. Anaphylaxis is very rare but does occur as with all vaccines. More severe adverse events, like Guillain-Barré syndrome have been reported with a particular vaccine in the 1970s but they are extremely rare. With the modern influenza vaccines, the causative risk is either found to be very rare (0.8 per million doses) [51] or there is no causal link found at all [52–54] and more association is found with influenza infection than vaccination [55]. An increased risk of fever and febrile seizures was reported from Australia in 2010 associated with the Southern Hemisphere IIV3 produced by CSL Biotherapeutics (now Sequirus) [39]. Following this, many countries do not recommend the CSL IIV3 for children aged <9 years.

Influenza vaccination during pregnancy has been shown to protect both the mother and her baby (up to 6 months old) against influenza [47, 48, 56, 57]. Influenza vaccination is safe in pregnancy and influenza vaccines have been administered to millions of pregnant women over many years and have not been shown to cause harm to pregnant women or their babies [58].

The most common adverse events associated with LAIV include runny nose or nasal congestion, headache, lethargy and sore throat. LAIV should not be administered to children with marked nasal congestion as this can reduce vaccine delivery. The safety of LAIV is not established in people with a history of asthma, diabetes mellitus, or other high-risk medical conditions associated with an elevated risk of complications from influenza. The use of LAIV in young children with chronic medical conditions is not recommended in the USA but is in some other countries.

2.7.7.1 Other Measures for Prevention of Influenza

Appropriate hand and respiratory hygiene (cough etiquette) has been shown to reduce influenza transmission in children in day-care or school [59]. Sick children and adults should remain at home and not attend school or work until symptoms have resolved to prevent transmission of influenza to others.

References

1. Nicholson KG, Wood JM, Zambon M. Influenza. Lancet. 2003;362(9397):1733–45.
2. Al-Tawfiq JA, Zumla A, Gautret P, Gray GC, Hui DS, Al-Rabeeah AA, et al. Surveillance for emerging respiratory viruses. Lancet Infect Dis. 2014;14(10):992–1000.
3. Hayward AC, Fragaszy EB, Bermingham A, Wang L, Copas A, Edmunds WJ, et al. Comparative community burden and severity of seasonal and pandemic influenza: results of the Flu Watch cohort study. Lancet Respir Med. 2014;2(6):445–54.
4. Nair H, Brooks WA, Katz M, Roca A, Berkley JA, Madhi SA, et al. Global burden of respiratory infections due to seasonal influenza in young children: a systematic review and meta-analysis. Lancet. 2011;378(9807):1917–30.
5. Lafond K, Nair H, Rasooly MH, Valente F, Booy R, Rahman M, et al. Global role and burden of influenza in pediatric respiratory hospitalizations, 1982–2012: a systematic analysis. PLoS Med. 2016;13(3):e1001977.

6. Simonsen L, Blackwelder WC, Reichert TA, Miller MA. Estimating deaths due to influenza and respiratory syncytial virus. JAMA. 2003;289(19):2499–500.
7. Fowlkes A, Steffens A, Temte J, Di Lonardo S, McHugh L, Lynfield R, et al. Incidence of medically attended influenza during pandemic and post-pandemic seasons through the Influenza Incidence Surveillance Project, 2009–13. Lancet Respir Med. 2015;3(9):709–18.
8. Cauchemez S, Ferguson N, Wachtel C, Tegnell A, Saour G, Duncan B, et al. Closure of schools during an influenza pandemic. Lancet Infect Dis. 2009;9(8):473–81.
9. Grohskopf L, Sokolow LZ, Broder K, Olsen SJ, Karron R, Jernigan D, et al. Prevention and control of seasonal influenza with vaccines recommendations of the Advisory Committee on Immunization Practices—United States, 2016–17 influenza season. Morb Mortal Wkly Rep. 2016;65(5):1–52.
10. Tempia S, Walaza S, Moyes J, Cohen A, Von Mollendorf C, Treurnich F, et al. Risk factors for influenza-associated severe acute respiratory illness hospitalization in a high HIV prevalence setting—South Africa, 2012–2015. Open Forum Infect Dis. 2017;4(1):ofw262.
11. Cohen C, Moyes J, Tempia S, Groom M, Walaza S, Pretorius M, et al. Severe influenza-associated respiratory infection in high HIV prevalence setting, South Africa, 2009-2011. Emerg Infect Dis. 2013;19(11):1766–74.
12. Cohen C, Moyes J, Tempia S, Groome M, Walaza S, Pretorius M, et al. Mortality amongst patients with influenza-associated severe acute respiratory illness, South Africa, 2009-2013. PLoS One. 2015;10(3):e0118884.
13. Fell DB, Savitz DA, Kramer MS, Gessner BD, Katz M, Knight M, et al. Maternal influenza and birth outcomes: systematic review of comparative studies. BJOG. 2017;124(1):48–59.
14. McAnerney JM, Cohen C, Moyes J, Besselaar TG, Buys A, Schoub BD, et al. Twenty-five years of outpatient influenza surveillance in South Africa, 1984-2008. J Infect Dis. 2012;206(Suppl 1):S153–8.
15. Hirve S, Newman LP, Paget J, Azziz-Baumgartner E, Fitzner J, Bhat N, et al. Influenza seasonality in the tropics and subtropics—when to vaccinate? PLoS One. 2016;11(4):e0153003.
16. Hirve S, Lambach P, Paget J, Vandemaele KA, Fitzner J, Zhang W. Seasonal influenza vaccine policy, use and effectiveness in the tropics and subtropics—a systematic literature review. Influenza Other Respi Viruses. 2016;10(4):254–67.
17. Simonsen L, Clarke MJ, Schonberger LB, Arden NH, Cox NJ, Fukuda K. Pandemic versus epidemic influenza mortality: a pattern of changing age distribution. J Infect Dis. 1998;178(1):53–60.
18. Simonsen L, Spreeuwenberg P, Lustig R, Taylor RJ, Fleming DM, Kroneman M, et al. Global mortality estimates for the 2009 influenza pandemic from the GLaMOR Project: a modeling study. PLoS Med. 2013;10(11):e1001558.
19. Brankston G, Gitterman L, Hirji Z, Lemieux C, Gardam M. Transmission of influenza A in human beings. Lancet Infect Dis. 2007;7(4):257–65.
20. Fiore A, Bridges C, Katz JM, Cox N. Inactivated influenza vaccines. In: Plotkin S, Orenstein W, Offit P, editors. Vaccines. Beijing: Elsevier; 2013. p. 257–93.
21. Poehling KA, Edwards KM, Weinberg GA, Szilagyi P, Staat MA, Iwane MK, et al. The under-recognized burden of influenza in young children. N Engl J Med. 2006;355(1):31–40.
22. Glezen WP. Clinical practice. Prevention and treatment of seasonal influenza. N Engl J Med. 2008;359(24):2579–85.
23. Livingston RA, Bernstein HH. Prevention of influenza in children. Infect Dis Clin North Am. 2015;29(4):597–615.
24. Selvennoinen H, Peltola V, Lehtinen P, Heikkinen T. Clinical presentation of influenza in unselected children treated as outpatients. Pediatr Infect Dis J. 2009;28:372–5.
25. Kumar S, Havens PL, Chusid MJ, Willoughby Jr RE, Simpson P, Henrickson KJ. Clinical and epidemiologic characteristics of children hospitalized with 2009 pandemic H1N1 influenza A infection. Pediatr Infect Dis J. 2010;29(7):591–4.
26. Jones BF, Stewart MA. Duration of cough in acute upper respiratory tract infections. Aust Fam Physician. 2002;31(10):971–3.
27. Eccles R. Understanding the symptoms of the common cold and influenza. Lancet Infect Dis. 2005;5(11):718–25.

28. Henderson FW, Collier AM, Sanyal MA, Watkins JM, Fairclough DL, Clyde Jr WA, et al. A longitudinal study of respiratory viruses and bacteria in the etiology of acute otitis media with effusion. N Engl J Med. 1982;306(23):1377–83.
29. Rihkanen H, Ronkko E, Nieminen T, Komsi KL, Raty R, Saxen H, et al. Respiratory viruses in laryngeal croup of young children. J Pediatr. 2008;152(5):661–5.
30. Newland JG, Laurich VM, Rosenquist AW, Heydon K, Licht DJ, Keren R, et al. Neurologic complications in children hospitalized with influenza: characteristics, incidence, and risk factors. J Pediatr. 2007;150(3):306–10.
31. Amin R, Ford-Jones E, Richardson SE, MacGregor D, Tellier R, Heurter H, et al. Acute childhood encephalitis and encephalopathy associated with influenza: a prospective 11-year review. Pediatr Infect Dis J. 2008;27(5):390–5.
32. Goenka A, Michael BD, Ledger E, Hart IJ, Absoud M, Chow G, et al. Neurological manifestations of influenza infection in children and adults: results of a National British Surveillance Study. Clin Infect Dis. 2014;58(6):775–84.
33. Agyeman P, Duppenthaler A, Heininger U, Aebi C. Influenza-associated myositis in children. Infection. 2004;32(4):199–203.
34. Harper SA, Bradley JS, Englund JA, File TM, Gravenstein S, Hayden FG, et al. Seasonal influenza in adults and children—diagnosis, treatment, chemoprophylaxis, and institutional outbreak management: clinical practice guidelines of the Infectious Diseases Society of America. Clin Infect Dis. 2009;48(8):1003–32.
35. CRDM. Healthcare workers handbook on influenza. The National Institute for Communicable Diseases (NICD); 2016.
36. Casscells SW, Granger E, Kress AM, Linton A. The association between oseltamivir use and adverse neuropsychiatric outcomes among TRICARE beneficiaries, ages 1 through 21 years diagnosed with influenza. Int J Adolesc Med Health. 2009;21(1):79–89.
37. Toovey S, Rayner C, Prinssen E, Chu T, Donner B, Thakrar B, et al. Assessment of neuropsychiatric adverse events in influenza patients treated with oseltamivir: a comprehensive review. Drug Saf. 2008;31(12):1097–114.
38. Hoffman KB, Demakas A, Erdman CB, Dimbil M, Doraiswamy PM. Neuropsychiatric adverse effects of oseltamivir in the FDA Adverse Event Reporting System, 1999-2012. BMJ. 2013;347:f4656.
39. Committee on Infectious Diseases. Recommendations for prevention and control of influenza in children, 2016-2017. Pediatrics. 2016;138.
40. Jefferson T, Jones MA, Doshi P, Del Mar CB, Hama R, Thompson MJ, et al. Neuraminidase inhibitors for preventing and treating influenza in healthy adults and children. Cochrane Database Syst Rev. 2014;4:CD008965.
41. World Health Organization. Vaccine against influenza WHO position paper—November 2012. Wkly Epidemiol Rec. 2012;47(87):461–76.
42. Vesikari T, Knuf M, Wutzler P, Karvonen A, Kieninger-Baum D, Schmitt HJ, et al. Oil-in-water emulsion adjuvant with influenza vaccine in young children. N Engl J Med. 2011;365(15):1406–16.
43. DiazGranados CA, Dunning AJ, Kimmel M, Kirby D, Treanor J, Collins A, et al. Efficacy of high-dose versus standard-dose influenza vaccine in older adults. N Engl J Med. 2014;371(7):635–45.
44. Madhi SA, Dittmer S, Kuwanda L, Venter M, Cassim H, Lazarus E, et al. Efficacy and immunogenicity of influenza vaccine in HIV-infected children: a randomized, double-blind, placebo controlled trial. AIDS. 2013;27(3):369–79.
45. Nohynek H, Baum U, Syrjanen R, Ikonen N, Sundman J, Jokinen J. Effectiveness of the live attenuated and the inactivated influenza vaccine in two-year-olds—a nationwide cohort study Finland, influenza season 2015/16. Euro Surveill. 2016;21(38).
46. Pebody R, Warburton F, Ellis J, Andrews N, Potts A, Cottrell S, et al. Effectiveness of seasonal influenza vaccine for adults and children in preventing laboratory-confirmed influenza in primary care in the United Kingdom: 2015/16 end-of-season results. Euro Surveill. 2016;21(38).

47. Madhi SA, Cutland CL, Kuwanda L, Weinberg A, Hugo A, Jones S, et al. Influenza vaccination of pregnant women and protection of their infants. N Engl J Med. 2014;371(10):918–31.
48. Zaman K, Roy E, Arifeen SE, Rahman M, Raqib R, Wilson E, et al. Effectiveness of maternal influenza immunization in mothers and infants. N Engl J Med. 2008;359(15):1555–64.
49. Eick AA, Uyeki TM, Klimov A, Hall H, Reid R, Santosham M, et al. Maternal influenza vaccination and effect on influenza virus infection in young infants. Arch Pediatr Adolesc Med. 2011;165(2):104–11.
50. Belongia EA, Simpson MD, King JP, Sundaram ME, Kelley NS, Osterholm MT, et al. Variable influenza vaccine effectiveness by subtype: a systematic review and meta-analysis of test-negative design studies. Lancet Infect Dis. 2016;16(8):942–51.
51. Lasky T, Terracciano GJ, Magder L, Koski CL, Ballesteros M, Nash D, et al. The Guillain-Barre syndrome and the 1992-1993 and 1993-1994 influenza vaccines. N Engl J Med. 1998;339(25):1797–802.
52. Hurwitz ES, Schonberger LB, Nelson DB, Holman RC. Guillain-Barre syndrome and the 1978-1979 influenza vaccine. N Engl J Med. 1981;304(26):1557–61.
53. Roscelli JD, Bass JW, Pang L. Guillain-Barre syndrome and influenza vaccination in the US Army, 1980-1988. Am J Epidemiol. 1991;133(9):952–5.
54. Kaplan JE, Katona P, Hurwitz ES, Schonberger LB. Guillain-Barre syndrome in the United States, 1979-1980 and 1980-1981. Lack of an association with influenza vaccination. JAMA. 1982;248(6):698–700.
55. Stowe J, Andrews N, Wise L, Miller E. Investigation of the temporal association of Guillain-Barre syndrome with influenza vaccine and influenza like illness using the United Kingdom General Practice Research Database. Am J Epidemiol. 2009;169(3):382–8.
56. Omer SB, Goodman D, Steinhoff MC, Rochat R, Klugman KP, Stoll BJ, et al. Maternal influenza immunization and reduced likelihood of prematurity and small for gestational age births: a retrospective cohort study. PLoS Med. 2011;8(5):e1000441.
57. Benowitz I, Esposito DB, Gracey KD, Shapiro ED, Vazquez M. Influenza vaccine given to pregnant women reduces hospitalization due to influenza in their infants. Clin Infect Dis. 2010;51(12):1355–61.
58. Tamma PD, Ault KA, del Rio C, Steinhoff MC, Halsey NA, Omer SB. Safety of influenza vaccination during pregnancy. Am J Obstet Gynecol. 2009;201(6):547–52.
59. Willmott M, Nicholson A, Busse H, MacArthur GJ, Brookes S, Campbell R. Effectiveness of hand hygiene interventions in reducing illness absence among children in educational settings: a systematic review and meta-analysis. Arch Dis Child. 2016;101(1):42–50.

HIV Infection

3

Raziya Bobat and Moherndran Archary

Abstract
This chapter is on the human immunodeficiency virus (HIV) and the specific effects of the virus on children. The field of HIV is a rapidly changing one as new research reveals new information in many aspects including the pathogenesis, prevention and in the field of antiretroviral therapy. This chapter strives to provide the latest updates.

The first part deals with the epidemiology, the structure and life cycle of the virus, the immunopathogenesis, modes of transmission and prevention thereof, and making the diagnosis of HIV in children. This is followed by the natural history of HIV infection, classification of HIV infection and immune status, and the clinical manifestations of the disease. The next section covers general management and antiretroviral therapy of the HIV-infected child and adolescent. The final section covers prevention including prevention of mother to child transmission and vaccines.

In the field of paediatric antiretroviral therapy (ART), there are continually changing guidelines, as new formulations and new drugs and new fixed-dose combinations become available. It is recommended that readers refer to the WHO and national guidelines for the latest guidelines.

R. Bobat (✉) • M. Archary
King Edward VIII Hospital, Durban, South Africa

Department of Paediatrics and Child Health, Nelson R Mandela School of Medicine,
University of KwaZulu-Natal, Durban, South Africa
e-mail: bobat@ukzn.ac.za

© Springer International Publishing AG 2017
R.J. Green (ed.), *Viral Infections in Children, Volume I*,
DOI 10.1007/978-3-319-54033-7_3

3.1   Introduction

Paediatric HIV infection was first described in the late 1980s. It has since become a largely preventable disease due to efforts to prevent mother to child transmission and has progressed from being almost universally fatal to being a chronic disease, thanks to early and accurate diagnosis, early initiation of treatment and improved antiretroviral therapy.

3.2   Epidemiology

According to UNAIDS reports, there were 36.7 million people living with HIV in 2015 [1, 2]. Of these, 1.8 million were children less than 15 years of age of whom the majority, 88%, lived in sub-Saharan Africa.

- There were 150,000 new HIV infections in children <15 years of age in 2015
- There were 600 new infections per day in children <15 years of age
- New HIV infections among children have declined by 50% since 2010
- 110,000 children (<15 years) died of AIDS in 2015
- HIV caused 410 child deaths every day
- Almost half, 49%, of all children living with HIV were accessing treatment in 2015, up from 21% in 2010
- Although mortality in children with HIV has decreased, there has been an upward trend in adolescents.

One of the main goals of the Global Plan was to reduce new HIV infections among children by 90%, from the baseline in 2009 (the benchmark year against which progress is measured) (UNAIDS 2015). At the end of 2014, Global Plan countries had reduced new infections among children by 48%. South Africa has made the greatest progress in reducing new pediatric infections by 76%. The current goal is ZERO new infections among children [3–8].

Among the 21 countries, 1.2 million pediatric infections have been averted since 1996 through the provision of antiretroviral medicines to pregnant women living with HIV.

Fewer HIV infections among children has also meant fewer acquired immuno deficiency syndrome (AIDS) related child deaths. Since 2000, AIDS-related mortality among children under the age of 5 has fallen by approximately 60%.

3.3   Viral Structure, Immunopathogenesis, and Life Cycle

3.3.1   The Virus

The human immunodeficiency virus is an RNA virus, belonging to the lentivirus family. Retroviruses are unable to replicate outside of living host cells. HIV virions contain a virus capsid, which consists of the major capsid proteins; the virion envelope, and the matrix protein (Fig. 3.1).

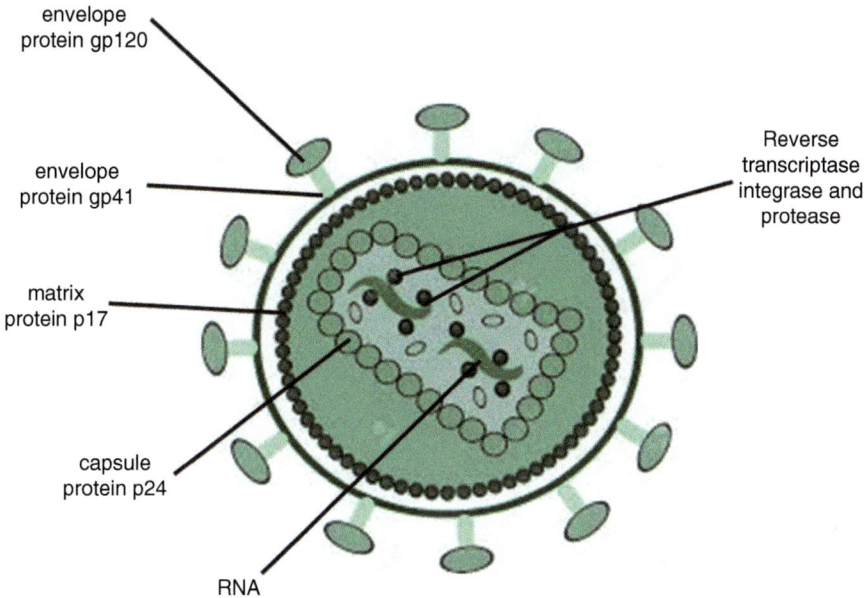

Fig. 3.1 Structure of a human immunodeficiency virus

There are two types of HIV that cause disease: HIV-1 and HIV-2. Most HIV-2 cases are found in western Africa and result in a less severe clinical disease spectrum. HIV-1 on the other hand has a worldwide distribution and is responsible for the HIV/AIDS pandemic. There are several subtypes of HIV-1, depending on the geographical area. The subtypes A and C are the most prevalent in South Africa.

HIV isolates show extensive genetic variability, with sequence variation even within the same patient. During the course of infection, the distribution of virus variants changes, providing the basis for the continuous emergence of new virus variants.

3.3.2 Immunopathogenesis

Infection with the HI virus results in a profound immunosuppression, making the host susceptible to various opportunistic infections and neoplasms. Effects of the virus on the host may be particularly dramatic in children as many of the organ systems are still developing.

The principal target of the virus is the CD4+ T lymphocyte, due to the presence of the CD4 molecule on the surface of the cell. In 1995, the first link between chemokines and the HI virus was established. CXCR4 and CCR5 were shown to serve as co-receptors for HIV on the cellular surface membrane. As gut epithelial cells do not express the CD4 receptor, the HI virus gains entry to the lamina propria by means of additional receptors and co-receptors. These co-receptors include CCR5 (used by virus that penetrates macrophages, e.g. in the gut) and CXCR4 (used by viruses that penetrate T-cells). Infection with HIV leads to destruction of the CD4 cells through direct or indirect effects. This loss of CD4 cells results in the characteristic immunosuppression.

A period exists between infection and manifestation of disease, during which viral replication occurs without any symptom manifestation. Numerous factors can induce HIV activation. These include: mitogens, cytokines, and soluble factors such as interferons and TGF-B.

Immunological changes in paediatric HIV infection include:

- T-cells:
  Total lymphopaenia
  CD4 lymphopaenia
  Reversal of CD4:CD8 ratio
- B-cells:
  Hypergamma-globulinemia (early abnormality)
  Abnormal antibody response to variety of antigens
  Polymorphonuclear neutrophils: Neutropenia
  Impaired chemotaxis and phagocytosis
  Chronic immune activation
  Increased apoptosis
  Failure of HIV-specific CD8 response (cytotoxic)
- Cytokine dysregulation:
  Production of proinflammatory cytokines
  Decreased production of IL-2 and IFN γ.

Eradication or cure of HIV following an established infection has proven difficult even with effective antiretroviral treatment due to the persistence of the virus in reservoirs and sanctuary sites [9, 10]. HIV can infect cells from the monocytic lineage (monocytes and macrophages), which express the CD4 molecule on their surface; these may serve as reservoirs for the virus. The virus persists indefinitely as an integrated part of the genome in a small population of latently infected cells. These latently infected cells remain relatively stable over time. Once reactivated, these latently infected cells can actively produce viral particles.

Soon after initial infection, the HI virus can enter sites of the body within which there is poor antiretroviral drug penetration and the virus can remain dormant. These are known as sanctuary sites. Areas of the body which can serve as sanctuary sites for HIV include: the CNS, lymph nodes, fat tissue, and the male genital tract. These reservoirs are also responsible for the re-emergence of the virus when ART is stopped even after many years on treatment. Studies have shown that the establishment of reservoirs in an infant can be markedly minimized if antiretroviral treatment is started very early in the newborn infant. Current research into a possible "cure" for HIV includes minimizing the amount of virus which enters sanctuary sites by early ART, developing drugs which can reach virus in the sanctuary sites, and getting the virus out of sanctuary sites so that it can be targeted by ART.

3.3.3 Life Cycle of the Human Immunodeficiency Virus

Knowing the life cycle of the virus is important in understanding the sites at which antiretroviral drugs act (Fig. 3.2 and see Sect. 3.8.6.2 for ART Drug Groups). There are several steps in the life cycle of the virus.

Step 1: *Attachment*
> The viral protein, GP120 attaches to CD4 receptors and co-receptors (CXCR4 and CCR5) on the surface of the host cell

Step 2: *Fusion*
> The cell membranes of the host cell and the envelope of the virus fuse; allowing entry of viral core (viral capsid with two copies of single stranded RNA with three enzymes, reverse transcriptase, integrase and protease) into the host cell

Step 3: *Reverse transcription*
> In the cytoplasm, reverse transcriptase, converts the viral single stranded RNA into a double-stranded DNA which can enter the nucleus of the host cell

Step 4: *Integration*
> Inside the nucleus of the host cell, the viral enzyme Integrase enables the viral DNA to be integrated with the host DNA

Step 5: *Replication*
> Following activation of the integrated viral DNA, messenger RNA is produced. mRNA uses the host's protein manufacturing machinery to produce new viral proteins in the cytoplasm.

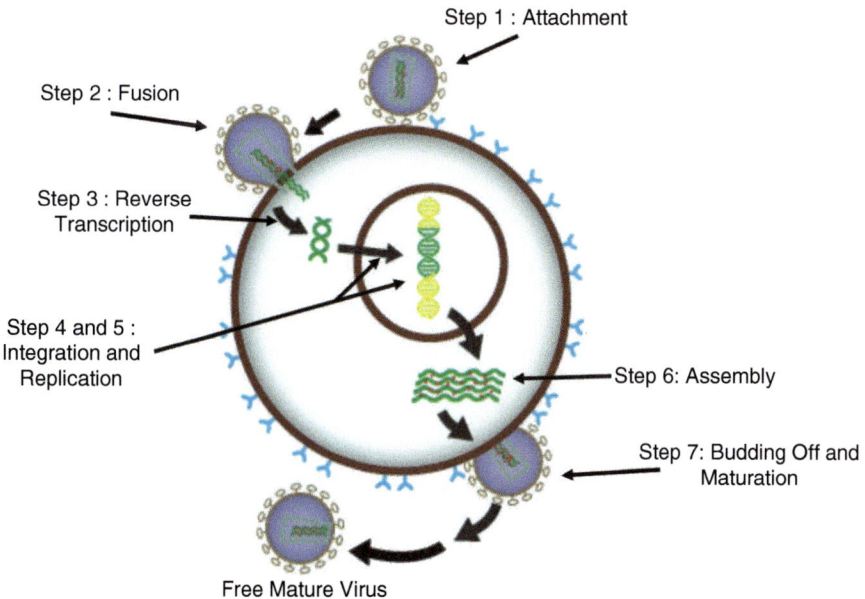

Fig. 3.2 HIV lifecycle

Step 6: *Assembly*

The viral proteins and RNA assemble at the cell membrane. The enzyme Protease cleaves the long proteins into appropriate sizes to form immature forms of the HI virus.

Step 7: *Budding off and maturation*

The new immature forms of the HI virus bud off from the host cell and mature into a new infectious virus.

3.4   Transmission

Most transmission in children (>90%) occurs by vertical transmission of the virus from an HIV-infected mother, either during pregnancy, during labour, or post natally through breastfeeding. The risk of vertical transmission has declined from as high as 30–45% to as low as less than 1% in most parts of the world.

There are many factors which increase the risk of transmission. These include:

- Maternal HIV stage
- Low maternal CD4 count
- High maternal viral load
- Acute infection of the mother during pregnancy (causing high level of viremia)
- Co-infections of the mother with syphilis, malaria, TB during pregnancy
- Premature/prolonged rupture of membranes
- Preterm delivery
- Vaginal delivery
- Mastitis during breastfeeding
- Mixed feeding (breast and formula).

Factors which have decreased the risk of transmission:

- ART during pregnancy for all women diagnosed with HIV
- Prompt treatment of sexually transmitted diseases during pregnancy
- Exclusive breastfeeding
- Administering ART to the infant during the period of breastfeeding
- Cesarean delivery
- Heat treatment of breast milk.

Other modes of transmission
Transmission of HIV can also occur by:

- sexual abuse of a child by an HIV-infected adult
- use of contaminated blood and blood products

- scarification by traditional healers, body piercing
- drug abuse
- In addition, transmission in adolescents can occur through unprotected sexual intercourse.

In children, of particular significance, is the relationship between viral load in the mother and transmission of HIV, as well as viral load and disease progression in the infant. A high viral load in the mother, or new infection during pregnancy, makes transmission and risk of infection more likely. Acute infection of the infant leads to a rapid increase in the viral load and causes a severe viremia. Viral loads tend to decrease slowly in vertically infected children and may not reach stable levels (viral set-point) until about 4–5 years of age.

3.5   Diagnosis

The diagnosis of HIV is an important gateway, facilitating access to the treatment cascade. This is of particular concern in infants and children, where delayed initiation of ART has been associated with a higher mortality and increased morbidity including neurodevelopmental delays [11]. Infants and children therefore require the identification of HIV infection before becoming clinically symptomatic and as soon as possible following acute infection. The strategy of HIV testing in infants involves routine screening of all infants born to HIV-infected mothers and early infant diagnosis (EID). In December 2015, UNAIDS launched it's 90:90:90 goal, i.e. by the year 2020, 90% of those infected must know their status, 90% of those infected must be on treatment, and 90% of those on treatment must be virologically suppressed [1, 2].

3.5.1 What Laboratory Tests to Perform?

3.5.1.1 HIV Virological Assays

The HIV Nucleic Acid Amplification Tests (NAAT) is a laboratory test using either whole blood/plasma or a dried blood spot (DBS), where HIV DNA and/or RNA in the sample is amplified and detected. A positive test indicates the presence of the HI virus in the patient's sample, which is extremely important in making the diagnosis of HIV in infants less than 18 months of age where passively acquired maternal preclude the use of an HIV serological test and following acute HIV infection before the production and detection of HIV antibodies by HIV serological tests. The value of the assay ranges from 95 to 98% sensitivity and 98% specificity. Sensitivity of the HIV PCR is affected by the quality of the sample, age of the infant and antiretroviral exposure.

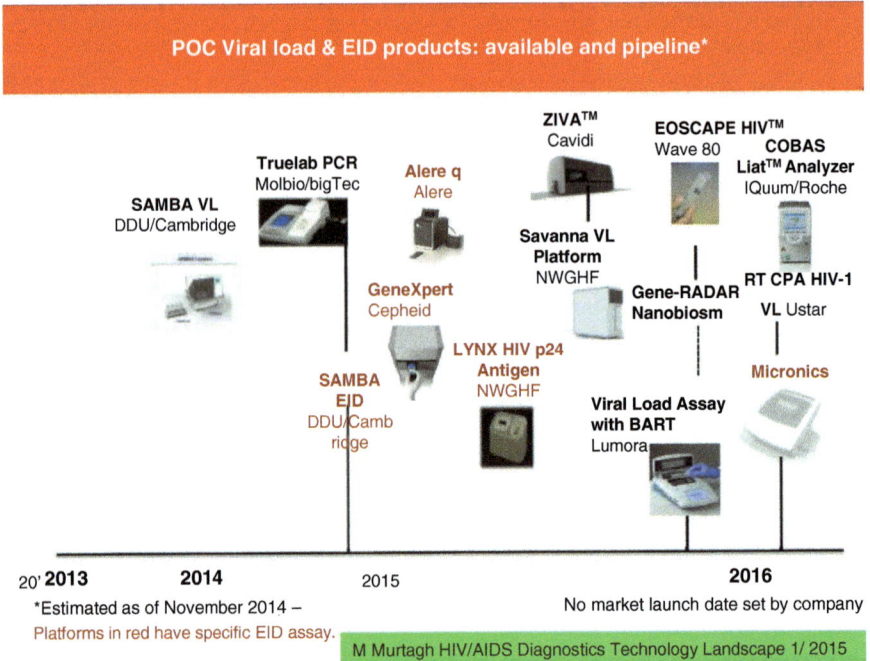

Fig. 3.3 Point-of-care HIV-related diagnostic tests for use in resource limited settings

3.5.1.2 Point of Care (POC) HIV Virological Assays

The advantage of POC testing is that it allows for the processing of samples at, or close to, the site of sample collection and decreases the turnaround time for HIV virological assays to a few hours rather than days or weeks. Several POC devices have been evaluated in field tests including the Alere q with a sensitivity of 95.5% (95% CI 91.7–97.9) and specificity of 99.8% (95% CI 99.1–100%). However the use of POC assays for routine birth testing still requires further evaluation (Fig. 3.3).

3.5.1.3 HIV Nucleic Acid Amplification Tests (NAAT)

HIV RNA NAAT

Both quantitative and qualitative HIV RNA NAAT assays are available for detection of extracellular HIV RNA for pediatric HIV diagnosis. HIV RNA viral loads can be affected by exposure to maternal antiretroviral treatment (ART) and infant antiretroviral prophylaxis regimens thus affecting the performance of the test. Ideally HIV RNA NAAT assays should be performed 2–4 weeks after stopping all ART exposure. Furthermore HIV RNA NAAT performed at birth will identify in-utero acquired HIV infections and not intra-partum acquired infections, therefore repeat testing at a later time point is required. However there is good correlation between the DNA and RNA NAAT assays, using a cut-off value of >5000 copies/mL. HIV RNA NAAT can be performed on dried blood spots (DBS) and plasma.

HIV DNA NAAT

Qualitative HIV DNA NAAT assays detect intracellular HIV DNA in peripheral blood mononuclear cells. HIV DNA NAAT assays have similar limitations as the RNA NAAT, in that when performed at birth, intra-partum HIV transmissions are not detected, therefore will only detect 20–55% of infected infants. The use of maternal ART to prevent intra-partum HIV transmission, has changed the epidemiology of neonatal HIV infections resulting in fewer HIV transmission which are more likely to be intra-uterine acquired. Furthermore HIV DNA NAAT assays are potentially affected to a lesser extent by maternal ART and infant antiretroviral prophylaxis than RNA DNA NAAT assay and are therefore the preferred assay for HIV birth testing.

3.5.1.4 HIV Antigen Tests (P24 Antigen)

Ultra-sensitive P24 antigen test detects the P24 antigen (Ag) (a viral core protein) in DBS and plasma samples with a 95% sensitivity and >99% specificity. When the HIV viral load is high, plasma concentrations of the p24 Ag are similarly high, making the test useful for early infant diagnosis and diagnosis of acute HIV infection during which the HIV viral load is particularly high. In settings where access to NAATs are not available, ultrasensitive P24 Ag tests are a valuable diagnostic tool to improve access to EID and treatment access for children. The test has not been evaluated for use for birth HIV testing.

3.5.1.5 HIV Antibody Tests

HIV antibody test detects HIV specific antibodies on capillary blood or serum, detecting antibodies against both HIV 1 and HIV 2. The test has a sensitivity of 99%, with a specificity of 98%. Third generation HIV antibody tests are able to detect concentrations of antibodies down to 2 µg/µL, and has a window period to 6–8 weeks from acute infection. The addition of P24 antigen in the fourth generation HIV antibody tests has decreased the window period to 4 weeks from acute infection. In children over 18 months of age, detection of HIV antibodies is diagnostic of HIV infection as maternal antibodies would have cleared. However, recent studies have demonstrated that with maternal ART, there may be persistence of maternal antibodies in the infants beyond 18 months. In a public health setting, failure to sero-revert (where a HIV-exposed infant reverts from a positive to a negative HIV antibody test) can be used as a cost-effective way of identifying an HIV positive infant where HIV NAAT testing is not available or cost-effective.

3.5.2 HIV Testing Strategies

See Table 3.1.

Table 3.1 Diagnostic strategies for infant HIV diagnosis

Age	WHO	US	PENTA	SA
Birth	Conditional recommendation (PCR)	Optional (PCR)	Yes (PCR)	Yes (PCR)
14–21 days		Yes (PCR)	Yes (PCR)	
4–6 weeks	Yes (PCR)			
Post cessation of infant prophylaxis		Yes (PCR) (at 4–6 weeks at least 2–4 weeks post cessation)	Yes (PCR) (at 2 and 6 weeks post cessation in high risk infants at 4 weeks)	Yes (PCR) at 10 weeks and/ or 18 weeks
6–9 months	Yes (Ab) if positive Ab test to do a HIV PCR	4–6 months (PCR)		
18 months	Yes (Ab)	Yes (Ab)	Yes (Ab)	Yes (Ab)
XX Symptomatic	Yes (PCR/Ab)	Yes (PCR/Ab)	Yes (PCR/Ab)	Yes (PCR/Ab)

PCR HIV PCR, *Ab* HIV serological test

3.6 Natural History and Classification

3.6.1 The Natural History of Paediatric HIV Infection

Without treatment, HIV infection causes progressive immunosuppression, due to HIV virus-mediated depletion of CD4+ lymphocytes, leaving patients at risk of developing opportunistic infections (OI) and other HIV-related disorders.

In the absence of ART and any form of prevention/intervention, three stages of disease progression may be recognized: rapid progression, intermediate progression and slow progression.

- Rapid progressors present by 3 months of life, progress rapidly, and die by 1–2 years of age
- Intermediate progressors have slower disease progression, tend to be admitted many times with varying infections, and usually progress to advanced disease by 5 years of age, and die by 5–6 years of age
- The slow progressors are those children who are vertically infected but for varying reasons do not progress from the infection until much later in life. They usually present to health services around 8–10 years of age, and often with TB or lung infection or cryptococcal disease.

Despite widespread access to prevention of mother to child transmission (PMTCT) and ART programmes, there are still far too many women who do not access treatment during pregnancy, and whose infants remain undiagnosed, and therefore do not access ART.

3.6.2 Classification of HIV Infection

Two classification systems are used in children, the WHO classification system, and the CDC classification system. Previously, the classification systems were used in making decisions about starting antiretroviral therapy. However, with the new treatment guidelines, these are used only for baseline assessment of newly diagnosed children and to assess disease progression or response to treatment (Tables 3.2, 3.3, 3.4).

Table 3.2 WHO staging of established HIV infection

Clinical stage 1
Asymptomatic
Persistent generalized lymphadenopathy
Clinical stage 2
Moderate unexplained weight loss
(<10% of presumed or measured body weight)[a]
Recurrent respiratory tract infections sinusitis, tonsillitis, otitis media and pharyngitis)
Herpes zoster
Angular cheilitis
Recurrent oral ulceration
Papular pruritic eruptions
Seborrhoeic dermatitis
Fungal nail infections
Clinical stage 3
Unexplained severe weight loss (>10% of presumed or measured body weight)
Unexplained chronic diarrhea for longer than 1 month
Unexplained persistent fever (above 37.6 °C (99.68 °F) intermittent or constant, for longer than 1 month)
Persistent oral candidiasis
Oral hairy leukoplakia
Pulmonary tuberculosis (current)
Severe bacterial infections (such as pneumonia, empyema, pyomyositis, bone or joint infection, meningitis or bacteremia)
Acute necrotizing ulcerative stomatitis, gingivitis or periodontitis
Unexplained anemia (neutropenia or chronic thrombocytopenia)
Clinical stage 4[b]
HIV wasting syndrome
Pneumocystis pneumonia
Recurrent severe bacterial pneumonia
Chronic herpes simplex infection (orolabial, genital or anorectal of more than 1 month's duration or visceral at any site)
Esophageal candidiasis (or candidiasis of trachea, bronchi or lungs)
Extrapulmonary tuberculosis
Kaposi's sarcoma
Cytomegalovirus infection (retinitis or infection of other organs)

(continued)

Table 3.2 (continued)

Central nervous system toxoplasmosis
HIV encephalopathy
Extrapulmonary cryptococcosis including meningitis
Disseminated non-tuberculous mycobacterial infection
Progressive multifocal leukoencephalopathy
Chronic cryptosporidiosis (with diarrhea)
Chronic isosporiasis
Disseminated mycosis (coccidiomycosis or histoplasmosis)
Recurrent non-typhoidal Salmonella bacteremia
Lymphoma (cerebral or B-cell non-Hodgkin) or other solid HIV-associated tumours Invasive cervical carcinoma
Atypical disseminated leishmaniasis
Symptomatic HIV-associated nephropathy or symptomatic HIV-associated cardiomyopathy

[a]Unexplained refers to where the condition is not explained by other causes
[b]Some additional specific conditions can also be included in regional classifications (such as reactivation of American trypanosomiasis [meningoencephalitis and/or myocarditis]) in the WHO Region of the Americas and disseminated penicilliosis in Asia)

Table 3.3 WHO immunological classification for established HIV infection

HIV-associated immunodeficiency	Age-related CD4 values			
	<11 months (%CD4+)	12–35 months (%CD4+)	36–59 months (%CD4+)	>5 years (absolute number per mm³ or %CD4+)
None or not significant	>35	>30	>25	>500
Mild	30–35	25–30	20–25	350–499
Advanced	25–29	20–24	15–19	200–349
Severe	<25	<20	<15	<200 or <15%

Table 3.4 CDC immunologic categories for HIV-infection in children based on absolute CD4$^+$ counts

Category	<1 y	1–5 y	6–12 y
1—No suppression	≥1500 (>25)	≥1000 (>25)	≥500 (>25)
2—Moderate suppression	750–1499 (15–24)	500–999 (15–24)	200–499 (15–24)
3—Severe suppression	<750 (<15)	<500 (<15)	<200 (<15)

3.7   Clinical Manifestations

Infants who are infected in-utero and do not receive ART, will manifest with signs/symptoms of HIV infection within 6 months of life. The following signs and symptoms may occur:

- Fever
- Skin rash
- Lymphadenopathy
- Hepatosplenomegaly
- Persistent diarrhea
- Recurrent infections (otitis, pneumonia)
- Oral candidiasis
- Delay in achieving developmental milestones
- Failure to thrive
- Parotidomegaly.

3.7.1 Systemic Manifestations/Effects of HIV Infection

3.7.1.1 Malnutrition and Growth Failure

Malnutrition is a common condition in HIV-infected children. Growth failure is a prominent manifestation of untreated HIV infection, with both stunting and wasting occurring early in infected children. Malnutrition and HIV together have an enhanced weakening of the immune system. Growth failure may be due to:

- poor appetite
- severe oro-pharyngeal candidiasis
- recurrent/persistent diarrhea
- malabsorption
- systemic infections such as tuberculosis
- opportunistic gut infections (isospora, cryptosporidium, mycobacterium avium complex (MAC))
- anemia
- poor health of the mother who may be unable to provide adequate food and supervision of the child's diet.

3.7.1.2 Skin

Skin lesions are extremely common in children with HIV infection, occurring in up to 90% of infected children, and may be the first manifestation of disease. Common dermatologic manifestations include fungal, bacterial, and viral infections of the skin; severe seborrheic dermatitis; vasculitis; and drug eruptions. HIV-infected children are additionally more likely to develop skin related side effects of drugs. Skin manifestations include:

- herpes simplex ulcers
- herpes zoster
- molluscum contagiosum
- warts
- candidiasis
- Kaposi sarcoma.

3.7.1.3 Hematological

HIV infection causes altered hematopoiesis can affect all cell lines (white blood cells, red blood cells, and platelets). The bone marrow in HIV-infected children undergoes changes, the most common being decreased cellularity and myelodysplasia. The commonest hematological abnormality is anemia, followed by thrombocytopenia, lymphopenia, and neutropenia.

Anemia in HIV-infected children is multi-factorial. Causes include:

- Direct effect of HIV on the marrow
- Iron deficiency
- Antiretroviral therapy
- Recurrent infections
- Parvovirus induced red cell aplasia
- Cytomegalovirus (CMV) infection.

The thrombocytopenia in HIV may be immune mediated, or due to depressive effect on the marrow, or due to opportunistic infections or the drugs used in their treatment.

There are no clear guidelines on management of the thrombocytopenia. ART should be initiated as early as possible. Response in the platelet count may be extremely slow, and in some children the thrombocytopenia remains refractory.

3.7.1.4 CNS and Development

HIV infection can have a profound effect on the developing brain. Manifestations of HIV encephalopathy may be due to the direct effects of HIV infection in the CNS, or to the effects of immune mediators. Neurodevelopmental and neurocognitive disorders are well-recognized complications of HIV infection. A large number of children with HIV infection will develop neurological involvement if untreated.

CNS manifestations include:

- HIV encephalopathy (triad of microcephaly, developmental delay or regression of milestones and tone disturbances)
- Seizures
- CNS infections (bacterial and cryptococcal meningitis, toxoplasmosis)
- Malignancy.

All HIV-infected should must have a developmental assessment done at first visit and regularly thereafter. Early use of ART may prevent deterioration, and may cause reversal of developmental delays.

3.7.1.5 Respiratory

Pulmonary disease is a frequent, and often the first, presentation in childhood HIV. It is associated with high morbidity and mortality. Recurrent pneumonia may be due to bacterial, viral or fungal organisms or mycobacteria. Multiple opportunistic pathogens may be present in infants presenting with severe pneumonia. Chronic lung disease is common in children with HIV who have not received early onset ART. Recent new research has shown that HIV affects the respiratory microbiome and this may be responsible for many of the lung abnormalities associated with HIV infection.

Pneumocystis jiroveci Pneumonia: Prior to early onset ART, *P. jiroveci* pneumonia (PCP) was responsible for approximately one-half of all AIDS-defining conditions diagnosed during the first year of life. Most infection occurred in the age range 3–12 months, with a peak at 3 months of age. Symptoms and signs include fever, tachypnea, cough, and hypoxia. On auscultation, there may be no additional sounds even in the presence of severe disease. A typical chest radiograph may show bilateral perihilar interstitial infiltrates.

Laboratory diagnosis is difficult and relies on microscopic demonstration of the organism. In resource-limited settings, appropriate specimens may be obtained through sputum induction or nasogastric aspirates. Alternately samples may be obtained from bronchoalveolar lavage where available.

The mortality rate in untreated infants is high. Treatment is initiated if PCP is suspected, especially if the mother is known to be HIV-infected. Treatment of PCP is with high dose cotrimoxazole, oxygen, and supportive care. The use of steroids is reserved for those with severe disease and or hypoxia.

Tuberculosis: In developing countries, tuberculosis (TB) is frequently found in children with HIV infection. There is extensive overlap between the clinical features of the two diseases and making the diagnosis is difficult. The tuberculin skin test is not useful due to the presence of immunosuppression, and the disease is often paucibacillary. Strenuous efforts should be made to diagnose or exclude TB in all HIV-infected children, and treatment initiated early. ART should be delayed for at least 2 weeks after starting TB treatment, to prevent immune reconstitution inflammatory syndrome (IRIS).

Lymphocytic interstitial pneumonitis (LIP): LIP is a chronic, progressive pulmonary disorder, usually presenting after the first year of life. Digital clubbing occurs as the disease progresses. Generalized lymphadenopathy and parotid and other salivary gland enlargement are usually present. A diffuse nodular pattern, with or without hilar or paratracheal lymph node enlargement, may be present on the chest X-ray. The condition responds well to HAART if initiated early. Long term complications of untreated LIP include bronchiectasis and pulmonary hypertension. With the rollout of early infant ART, the number of children with LIP has dramatically decreased.

3.7.1.6 GIT

Gastrointestinal symptoms including anorexia, vomiting and diarrhea are extremely common. In the developing world, these occur against the background of a very high incidence of acute infective diarrhea, which may be viral or bacterial. There may be acute diarrhea with dehydration, or persistent diarrhea with failure to thrive. Combinations of malabsorptive features may be found including lactose or monosaccharide malabsorption. HIV-infected children with severe acute malnutrition (SAM) who develop diarrhea have a higher risk of developing serious complications from their diarrhea and have much worse outcomes. These children should be admitted and monitored carefully.

Opportunistic intestinal infections occur more frequently in those with severe immunocompromise, and include Cryptosporidium, Isospora and Giardia. In children with severe failure to thrive due to recurrent diarrhea, admission and total parenteral nutrition may be required before any weight gain is noted.

Research has shown that HIV affects the microbiome of the GIT, and this may be responsible for the numerous problems occurring in the gut, including diarrhea, malabsorption, and opportunistic infections.

3.7.1.7 CVS
HIV-associated cardiomyopathy, pericardial effusion, and conduction abnormalities have been reported in children. These generally manifest later in childhood. Pulmonary hypertension may occur in children with chronic lung disease. In addition, cardiovascular abnormalities are now being seen in older children on long term ART.

3.7.1.8 Renal
HIV infection may cause several types of renal disease. The most common clinical entity in children is HIV-associated nephropathy (HIVAN). Other renal manifestations of HIV infection are glomerulonephritis and nephrotic syndrome, although less common. Patients with HIVAN should be treated with ART, which decreases the rate at which HIVAN progresses to renal failure.

3.7.1.9 Malignancies
Children with HIV infection are at increased risk of developing malignancies. The most common malignancies found in children with HIV are lymphoma and Kaposi sarcoma. The prevalence of HIV-related malignancies has been declining with the use of ART. Results from a recent study revealed that those who received treatment immediately on diagnosis had a much lower risk of developing Kaposi sarcoma and lymphoma [12].

3.7.1.10 Opportunistic Infections
One of the major effects of the immunosuppression associated with HIV infection is the predilection for OI's. Some of the OI's that occur in children in developing countries include TB, disseminated candidiasis, herpes virus infections, cryptococcal disease, CMV infection, MAC, and intestinal organisms such as cryptosporidium. Transmission of OI's in children can occur either vertically from the mother; or horizontally from the environment. It is possible that HIV-infected women are more likely to transmit opportunistic pathogens to their infants. Generally, an OI in a child is a primary infection (as opposed to an adult where it is usually a re-activation). The risk of an OI increases as the CD4 count declines. Over the past few years, much has

occurred to change the landscape of OI's in children. These changes include the following:

- Early initiation of HAART
- Revised criteria for primary and secondary prevention
- New diagnostic tests for OI's
- New vaccines becoming available e.g. HPV, pneumococcal
- The development of new drugs for the treatment of OI's, and the development of pediatric formulations.

An international team conducted a meta-analysis of the prevalence of OI's and found that in low and middle income countries, there has been a major decline in the number of cases of new opportunistic infections among children receiving ART [13].

3.8   Management

3.8.1   General

Comprehensive care of the HIV-infected child includes the following:

- Counseling on infant feeding and infant feeding choices and support thereof
- Good nutrition, nutritional support, supplements
- Growth monitoring
- Routine immunization
- Routine treatment for intestinal helminths
- Support of the care giver
- Regular counseling of the care giver
- Prophylaxis as per the national guidelines
- Antiretroviral therapy
- Monitoring of treatment response and side effects.

3.8.2 Prophylaxis

3.8.2.1 Co-trimoxazole
- Co-trimoxazole prophylaxis is recommended for HIV-exposed infants from 4 to 6 weeks of age and should be continued until HIV infection has been excluded by an age-appropriate HIV test to establish final diagnosis after complete cessation of breastfeeding
- Co-trimoxazole prophylaxis is recommended for infants, children, and adolescents with HIV, irrespective of clinical and immune conditions. Priority should be given to all children less than 5 years old regardless of CD4 cell count or clinical stage, and children with severe or advanced HIV clinical disease (WHO clinical stage 3 or 4) and/or those with CD4 $\leq$ 350 cells/mm^3.

3.8.2.2 Tuberculosis

- Children living with HIV who have poor weight gain, fever or current cough or contact history with a TB case may have TB and should be evaluated for TB and other conditions. If the evaluation shows no TB, they should be offered IPT preventive therapy regardless of their age.

3.8.2.3 Cryptococcus (Cr)

- The routine use of antifungal primary prophylaxis for cryptococcal disease in HIV-infected adults, adolescents and children with a CD4 count less than 100 cells/mm^3, and who are CrAg-negative or where CrAg status is unknown, is not recommended prior to ART initiation.

3.8.3 Infant Feeding

The new WHO guidelines on infant feeding [3–8] recommend:
Factors that may decrease the risk of HIV transmission through breastfeeding include:

- Shorter duration of breastfeeding. The longer a child is breastfed by an HIV-positive mother the higher the risk of HIV infection
- Exclusive breastfeeding in the early months. Exclusive breastfeeding during the first 3 months of life results in a lower risk of MTCT than mixed feeding
- Prevention and treatment of breast problems. Mastitis and cracked nipples and other causes of breast inflammation are associated with an increased risk of HIV-transmission
- Prevention of HIV-infection during breastfeeding. The maternal viral load is higher shortly after a new infection resulting in an increased risk of infection of the child
- Early treatment of sores or thrush in the mouth of the infant. Sores in the infant's mouth make it easier for the virus to enter the infant's body
- Maternal and infant ART: antiretroviral therapy taken by mother during pregnancy and for the duration of breastfeeding as well as ART taken by the infant for the first 12 weeks of life.

3.8.4 Nutrition

Early nutritional intervention (i.e. nutritional assessment and support) should be an integral part of the care plan of HIV-infected children. In asymptomatic HIV-infected children, resting energy expenditure is increased by about 10%, while increases in energy needs of between 50 and 100% have been reported in HIV-infected children experiencing growth failure.

It is recommended to increase the energy intake of HIV-infected infants and children by 10% of the RDA for their age and sex if they are asymptomatic, and by 20–30% of the RDA if they are symptomatic or recovering from acute infections [3–8] (Table 3.5).

Table 3.5 Total energy needs of HIV-infected children (kcal/day)

	Daily energy needs of HIV-infected children[a]	HIV-infected and asymptomatic 10% additional energy	HIV-infected and poor weight gain or other symptoms 20% additional energy	Severely malnourished and HIV-infected (post stablilization) 50–100% additional energy[b]
6–11 months	690	760	830	150–220 kcal/kg/day
12–23 months	900	990	1080	150–220 kcal/kg/day
2–5 years	1250	1390	1510	150–220 kcal/kg/day
6–9 years	1650	1815	1980	75–100 kcal/kg/day
10–14 years	2020	2220	2420	60–90 kcal/kg/day

[a]Based on average of total energy requirements for light and moderate habitual physical activity levels for girls and boys by age group. Joint FAO/WHO/UNU Expert Consultation, October 2001. ftp://ftp.fao.org/docrep/fao/007/y5686e/y5686e00.pdf
[b]Management of Severe Malnutrition: a manual for physicians and other senior health workers 14

Children with SAM and any medical complications, or if below 6 months of age must be hospitalized. HIV-positive children with SAM should be receive urgent treatment including daily assessment. They should be nursed in a high care area until they are feeding well, infections are under control and diarrhea has stopped. Treatment is aimed at managing the following serious complications: hypoglycemia, hypothermia, dehydration, electrolyte imbalances, micronutrient deficiencies and infections.

Vitamin A supplementation as per national guidelines as well as regular deworming is important components of care.

3.8.5 Immunization

HIV-infected children are at high risk of developing severe disease from vaccine preventable infections. Therefore it is imperative to ensure that all HIV-infected children receive routine immunizations as per the country schedule. Although some of the vaccines are live attenuated (measles, rotavirus, BCG) their protective effects probably outweigh any potential side effects. It is, however, recommended that children with severe immune compromise should not receive the measles and varicella vaccines. The WHO recommends avoiding BCG at birth in HIV-infected infants, but this may be difficult in areas where TB and HIV are major health problems, and where there may be limited access to early diagnosis of HIV as well as risk of loss to follow up of newborns. Therefore each country needs to decide on an immunization schedule that is appropriate for it. In South Africa (SA), BCG is still given at birth.

SA has moved to the IPV vaccine in addition to two oral doses given at birth and 6 weeks. Measles vaccine has been revised as of December 2015, so that measles vaccine is given at 6 months and 12 months.

3.8.6 Antiretroviral Treatment

The use of combination ART has changed the natural history of HIV infection from a universally fatal condition to a chronic disease. Early treatment strategies including zidovudine (AZT) mono-therapy and dual therapy made way for the current standard three-drug combination therapy, which provides durable viral suppression when treatment compliance is maintained above 95% and minimizes drug related adverse effects.

The field of paediatric HIV is constantly evolving with new antiretroviral drugs and better formulations of older drugs providing more options for the treatment of HIV-infected children that are safer, easier to administer, more palatable and with a higher genetic barrier against the development of resistance. The goal of ART in children is to maintain an undetectable HIV viral load with a normal CD4 count and good clinical response (including growth and neurodevelopment). The second and third 90 in the WHO 90:90:90 strategy related to the goal of ensuring that 90% of eligible HIV-infected individuals start ART and 90% on patients start ART must be virologically suppressed [1, 2]. Table 3.6 provides an overview of ART management in naive children.

3.8.6.1 When to Start?
ART initiation criteria have evolved as the pediatric HIV epidemic has changed over time. Currently ART is a chronic treatment strategy, which requires life-long treatment. The benefits of immediate initiation of ART must be weighed against the potential harm from adverse effects of long term ART use and difficulties in maintaining good adherence throughout childhood into adolescence and adulthood.

The CHER study [11] clearly demonstrated a lower mortality and long-term neurocognitive benefit in early ART initiation before immunological and clinical deterioration in infants under the age of 1 year. In older children this clinical benefit has not been demonstrated; however availability of safer, easier to administer formulations, extrapolated data from adult studies demonstrating benefit of early ART initiation and programmatic advantages of a simplified treatment cascade has moved the pediatric ART field towards universal treatment. Currently the WHO, US, PENTA and South African ART treatment guidelines advocate initiation of ART in all HIV-infected children and adolescents on diagnosis of HIV irrespective of CD4 count, viral load or clinical stage.

3.8.6.2 What to Start?
Availability of suitably formulated antiretroviral drugs for use in a specific age category is a major deciding factor in determining the appropriate regimen. In the neonatal period the choice of available drugs are severely limited due to the lack of drugs that have been studied in this age group. Similarly the ability of an infant to swallow solid tablet formulations or tolerate unpalatable liquid formulations may also limit treatment choices. However the principles of treatment remain the same and the use of sub-optimal regimens should never occur.

Table 3.6 Overall management of an ART naïve patient

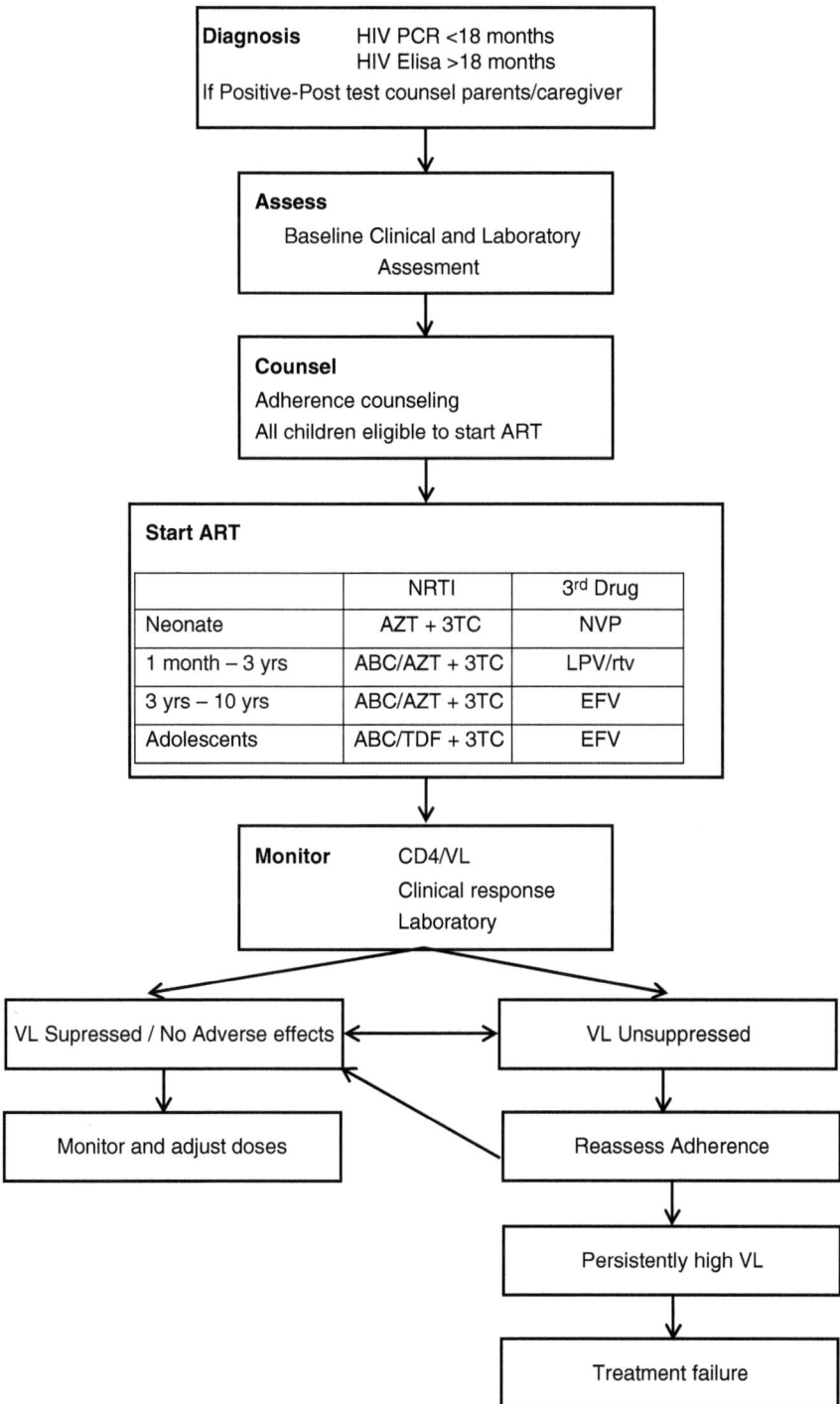

```
┌─────────────────────────────────────────────────┐
│ Diagnosis      HIV PCR <18 months                 │
│                HIV Elisa >18 months               │
│   If Positive-Post test counsel parents/caregiver │
└─────────────────────────────────────────────────┘
                        │
                        ▼
        ┌─────────────────────────────────────┐
        │ Assess                               │
        │    Baseline Clinical and Laboratory  │
        │            Assesment                 │
        └─────────────────────────────────────┘
                        │
                        ▼
        ┌─────────────────────────────────────┐
        │ Counsel                              │
        │ Adherence counseling                 │
        │ All children eligible to start ART   │
        └─────────────────────────────────────┘
                        │
                        ▼
```

Start ART

	NRTI	3rd Drug
Neonate	AZT + 3TC	NVP
1 month – 3 yrs	ABC/AZT + 3TC	LPV/rtv
3 yrs – 10 yrs	ABC/AZT + 3TC	EFV
Adolescents	ABC/TDF + 3TC	EFV

```
                        │
                        ▼
        ┌─────────────────────────────────────┐
        │ Monitor      CD4/VL                  │
        │              Clinical response       │
        │              Laboratory              │
        └─────────────────────────────────────┘
              │                      │
              ▼                      ▼
┌──────────────────────────┐  ┌──────────────────────┐
│ VL Supressed / No        │◄─►│ VL Unsuppressed      │
│ Adverse effects          │  │                      │
└──────────────────────────┘  └──────────────────────┘
              │                      │
              ▼                      ▼
┌──────────────────────────┐  ┌──────────────────────┐
│ Monitor and adjust doses │  │ Reassess Adherence   │
└──────────────────────────┘  └──────────────────────┘
                                     │
                                     ▼
                              ┌──────────────────────┐
                              │ Persistently high VL │
                              └──────────────────────┘
                                     │
                                     ▼
                              ┌──────────────────────┐
                              │ Treatment failure    │
                              └──────────────────────┘
```

Combination ART should consist of three or more active ART agents, from at least two different classes of drugs. Table 3.7 provides a summary of drugs currently registered for use in children.

Table 3.7 Drugs table

Group	Drug	Registration age (FDA approval)	Significant adverse effects
Attachment inhibitors	Investigational drugs (GP120 blocker: fostemsavir)		
Entry/ fusion inhibitors	Maraviroc (MVC)	>16 years	Hepatitis
			Rash
	Enfuvitide (T-20)	>6 years	Local injection site reactions
			Hypersensitivity reactions
RTI	Abacavir (ABC)	>3 months	Hypersensitivity reaction (fever, rash with multiorgan dysfunction) associated with HLA B5701—uncommon in children of African decent
	Lamivudine (3TC)	4 weeks (treatment)	Well tolerated
		Birth (prophylaxis)	
	Emtricitabine (FTC)	Birth	Well tolerated
	Zidovudine (AZT)	6 weeks (treatment)	Cytopenia (macrocytic anemia, neutropenia)
		Birth (prophylaxsis)	Gastrointerstinal effects
			Lactic acidosis
			Myopathy
			Lipodystrophy
			Metabolic effects
	Stavudine (D4T)	Birth	Peripheral neuropathy
			Lactic acidosis
			Pancreatitis
			Lipodystrophy
			Metabolic effects
	Didanosine (ddi)	> 2 weeks	Gastrointerstinal effects
			Peripheral neuropathy
			Lactic acidosis
			Pancreatitis
NtRTI	Tenofovir (TDF)	>2 years (oral powder)	Gastrointerstinal effects
			Renal effect (renal failure or renal tubulopathy)
			Bone effect (decreased bone mineral density)
	Tenofovir alafenamide (TAF)	>18 years	Lower renal and bone effect compared to TDF

Table 3.7 (continued)

Group	Drug	Registration age (FDA approval)	Significant adverse effects
NNRT	Efavirenz (EFV)	>3 years	Rash
			CNS effects (insomnia, abnormal dreams, hallucinations)
			Increased liver enzymes
			Lipodystrophy
			Gynecomastia
	Nevirapine (NVP)	2 weeks (treatment)	Rash (Stevens-Johnson syndrome)
		Birth (prophylaxis)	Hepatitis
	Etravirine (ETR)	>6 years	Rash (Stevens-Johnson syndrome)
			Hypersensitivity reaction
	Rilpivirine (RPV)	>18 years	CNS effects (mood disturbances, insomnia, headaches)
			Rash
	Dorovirine	Investigational drug active against EFV/RPV resistant virus	
INSTI	Raltegravir	>4 weeks	Rash
			Gastrointerstinal effects
			Myopathy
	Dolutegravir	>6 years	CNS effects (insomnia, suicidal ideation)
	Elvitegravir (EVG)	>18 years	Gastrointerstinal effects
			Renal effects
			Bone effect (decreased bone mineral density
	Cabotegravir	Investigational long acting injectable ART	
PI	LPV/rtv	>2 weeks and >42 weeks corrected gestational age	Gastrointerstinal effects
			Metabolic effects
			Lipodystrophy
	Atazanavir (ATV)	>6 years	Jaundice
			ECG abnormalities
			Metabolic effects
			Rash
			Lipodystrophy
	Darunavir (DRV)	>3 years/>10 kg	Rash (SJS)
			Hepatitis
			Metabolic effects
			Lipodystrophy
	Fosamprenavir (FPV)	>4 weeks	Gastrointerstinal effects
			Rash
			Metabolic effects
			Lipodystrophy
Maturation inhibitors	Investigational drugs (first generation: bevirimat, second generation: BMS-955176, GSK 2828232)		

The most common first-line combination therapy consists of: two nucleoside reverse transcriptase inhibitors (NRTI) with one drug from another class (either a non-nucleoside reverse transcriptase inhibitor (NNRTI), protease inhibitor (PI) or an integrase inhibitor (INSTI)). Table 3.8 summarizes the current treatment guidelines.

The use of fixed drug combinations (FDCs) and once a day regimens is highly recommended due to the ease of use for the patients and improved treatment adherence. Table 3.9 summarizes the FDCs currently available and in the developmental pipeline.

In adult patients, various treatment strategies have been attempted to simplify the ART regimen once a patient is virally suppressed. These include NRTI sparing regimens, PI mono-therapy, weekends off treatment and long acting injectable treatment at 3 monthly intervals. These interventions have not been evaluated in the pediatric population; however they may be part of future treatment strategies.

Table 3.8 First line regimens in children

Age category	WHO[5]	US Guidelines[18, 20]	PENTA [15, 16, 22]	SA[17]
Neonates	<2 weeks: AZT + 3TC + NVP	AZT + 3TC + NVP or AZT + 3TC + LPV/rtv (≥14 days)		<4 weeks: AZT + 3TC + NVP
	2 weeks–3 months: ABC/AZT + 3TC + LPV			>4 weeks: ABC + 3TC + LPV/rtv
1 month – 3 years	ABC/AZT + 3TC and LPV/rtv	AZT/ABC + 3TC and LPV/rtv or RAL	ABC/AZT + 3TC and LPV/rtv or NVP	ABC + 3TC and LPV/rtv
>3 years	ABC + 3TC and EFV	AZT/ABC + 3TC and ATV/rtv or DRV/rtv or EFV or LPV/rtv or RAL	ABC + 3TC and EFV or LPV/rtv	ABC + 3TC and EFV
Adolescents	TDF + 3TC/FTC and EFV	ABC/TAF + 3TC/FTC and ATV/rtv or DTG or DRV/rtv or EVG/cobistat	ABC/TDF + 3TC/FTC and EFV	<15 years: ADC + 3TC + EFV
				>15 years: TDF + 3TC + EFV

Table 3.9 Fixed drug combinations (FDCs) for use in children <18 years of age

Drugs	Doses (mg)	Formulations	Approval status
TDF/FTC/EFV	300/200/600	Tablet	Approved
ABC/3TC/EFV	150/75/150	Tablet	In development
AZT/3TC	60/30	Tablet	Approved
AZT/3TC/NVP	60/30/50	Dispersible tablet	Approved
ABC/3TC	60/30 and 120/60	Dispersible tablet	Approved
ABC/3TC/LPV/rtv	30/15/40/10	Granule	In development
AZT/3TC/LPV/rtv	30/15/40/10	Granule	In development
TDF/FTC/EVG/COBI		Tablet	In development
TAF/FTC/EVG/COBI		Tablet	In development
ABC/3TC/DTG		Dispersible tablet	In development

3.8.6.3 How to Monitor?

Following initiation of ART, the patient requires regular evaluation to assess for drug-related adverse effects and for an appropriate response to treatment. Patients who do not achieve an appropriate response to ART after at least 6 months on an appropriate regimen are considered to have treatment failure. In very young infants with high baseline HIV viral loads, it is acceptable to wait for at least 12 months on ART for viral suppression.

Table 3.10 provides the WHO definitions of clinical, immunological and virological failure.

Table 3.10 WHO definitions of clinical, immunological and virological failure for the decision to switch to ART regimens

Failure	Definition	Comments
Clinical failure	*Adults and adolescents*	The condition must be differentiated from IRIS occurring after initiating ART
	New or recurrent clinical event indicating severe immunodeficiency (WHO clinical stage 4 condition) after 6 months of effective treatment	For adults, certain WHO clinical stage 3 conditions (pulmonary TB and severe bacterial infections) may also indicate treatment failure
	Children	
	New or recurrent clinical event indicating advanced or severe immunodeficiency (WHO clinical stage 3 and 4 clinical condition with exception of TB) after 6 months of effective treatment	
Immunological failure	*Adults and adolescents*	Without concomitant or recent infection to cause a transient decline in the CD4 cell count
	CD4 count falls to the baseline (or below) or persistent CD4 levels below 100 cells/mm^3	A systematic review found that current WHO clinical and immunological criteria have low sensitivity and positive predictive value for identifying individuals with virological failure. The predicted value would be expected to be even lower with earlier ART initiation and treatment failure at higher CD4 cell counts. There is currently no proposed alternative definition of treatment failure and no validated alternative definition of immunological failure
	Children	
	Younger than 5 years: persistent CD4 levels below 200 cells/mm^3 or <10%	
	Older than 5 years: persistent CD4 levels below 100 cells/mm^3	

(continued)

Table 3.10 (continued)

Failure	Definition	Comments
Virological failure	Plasma viral load above 1000 copies/mL based on two consecutive viral load measurements after 3 months, with adherence support	The optimal threshold for defining virological failure and the need for switching ART regimen has not been determined
		An individual must be taking ART for at least 6 months before it can be determined that a regimen has failed assessment of viral load using DBS and point-of-care technologies should use a higher threshold

Early identification of treatment failure is extremely important in order to identify the likely contributing factors and intervening before the emergence of drug resistance. Re-suppression of the viral load is often possible if adherence is improved. However patients with persistent treatment failure require an evaluation, which may include an HIV drug-resistance test and/or a change in regimen.

3.8.6.4 HIV Drug Resistance (DR) Testing

Phenotypic and genotypic DR testing can be performed to detect resistance of the HIV to the patients' ART regimen. In clinical practice genotypic DR testing is most frequently performed due to the advantages in the cost and turnaround time.

Genotypic DR testing involves sequencing of specific regions of the HIV genetic sequences that code for ART target sites. These sequences are compared to a reference "wild-type" sequence and differences are detected as mutations. Only mutations present in over 20–25% of the viral population are detected using routine genotypic DR testing although deep and ultra-deep sequencing DR testing can detect mutations present in over 10% and 2% of the viral population, respectively. Genotypic DR testing does not detect archived mutations and may not detect mutations acquired during previous ART exposure. Therefore a complete and careful ART drug history including infant prophylaxis is required in order to interpret the DR testing. In addition patients must be on ART at the time of the DR testing.

Mutations may result in reduced susceptibility to one drug but increased susceptibility to another or require a combination of mutations for the virus to be resistant to a drug. Software such as the Standford database is available to simulate the effects of multiple mutations on antiretroviral drugs using standard algorithms and provides the clinician with a "virtual phenotypic DR test". The software classifies each drug as fully susceptible, low level, potential low-level, intermediate and high-level resistance similar to an antibiotic sensitivity test.

Selection of an appropriate regimen following a DR test requires an understanding of the drug susceptibility pattern and the effect of prior ART drug exposure, together with availability of drug formulations for the patient.

3.8.6.5 Special Issues with Treatment

Adolescents
The WHO definition of adolescence is an individual between the ages of 10 and 19 years. During this time of transition into adulthood, there are physical, endocrine and psychological changes. Adolescents acquire HIV either through perinatal acquisition (perinatally infected) or through sexual exposure (behaviorally infected).

Perinatally infected adolescents tend to be highly ART treatment experienced, and often have clinical consequences of chronic HIV infection including growth disturbances (short stature), delays in pubertal development, sequela of opportunistic infections, and HIV-associated organ dysfunction. Disclosure of their HIV status and dealing with the mode of transmission (infected by a parent) may result in conflict and resentment if not managed appropriately. They also have to navigate disclosure of their HIV status to people outside of their immediate family as their circle of influence increases.

Adolescence is also a period of self-discovery and asserting of one's independence, which is often manifested as rebellion against authority. This in combination with pill fatigue may be present with poor adherence to ART or refusal to continue ART.

Mental health issues are common in HIV-infected adolescents. Common clinical conditions include depression, anxiety disorders and attention deficit disorders.

Awareness of the potential issues facing perinatally infected adolescents is important for early identification and management.

Behaviorally-infected adolescents on the other hand, are sexually mature and are often without prior medical conditions. Their clinical management is often similar to management of an adult patient with HIV infection.

All adolescents irrespective of routine of acquisition benefit from receiving services that are tailored to the needs of adolescents i.e. youth friendly services (YFS). Some of the key characteristics of YFS are:

- Adolescent engagement during the development of the services at the facility
- Training of health care workers to engage with adolescents—Non-judgmental, easy to relate to, treat adolescents with respect
- Convenient clinic hour (after-school, over weekends, less frequent visits if stable)
- Reproductive and sexual health services (contraception/STI Treatment)
- Conducive environment for adolescents
- Youth lead support groups.

3.8.6.6 Immune Reconstitution Inflammatory Syndrome (IRIS)
IRIS is characterized by an excessive inflammatory response to an endogenous or exogenous antigen resulting in a paradoxical clinical worsening following ART initiation and immune restoration (Table 3.11).

Table 3.11 IRIS

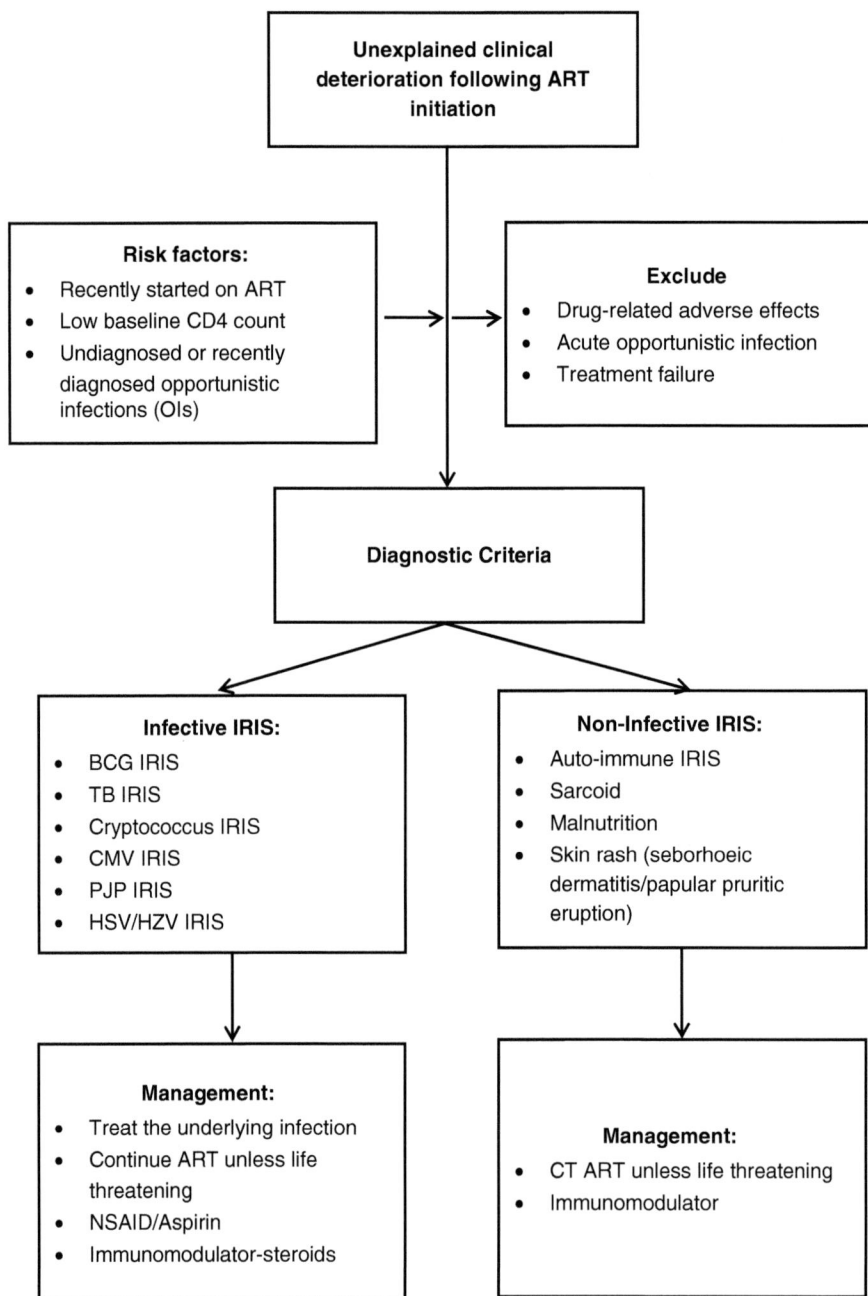

```
┌─────────────────────────────────┐
│   Unexplained clinical          │
│   deterioration following ART   │
│   initiation                    │
└─────────────────────────────────┘
```

Risk factors:
- Recently started on ART
- Low baseline CD4 count
- Undiagnosed or recently diagnosed opportunistic infections (OIs)

Exclude
- Drug-related adverse effects
- Acute opportunistic infection
- Treatment failure

Diagnostic Criteria

Infective IRIS:
- BCG IRIS
- TB IRIS
- Cryptococcus IRIS
- CMV IRIS
- PJP IRIS
- HSV/HZV IRIS

Non-Infective IRIS:
- Auto-immune IRIS
- Sarcoid
- Malnutrition
- Skin rash (seborhoeic dermatitis/papular pruritic eruption)

Management:
- Treat the underlying infection
- Continue ART unless life threatening
- NSAID/Aspirin
- Immunomodulator-steroids

Management:
- CT ART unless life threatening
- Immunomodulator

Risk factors for the development of IRIS include low CD4 count at ART initiation with a robust immune or viral response to ART. IRIS is a diagnosis of exclusion and other clinical conditions such as drug reactions, treatment failure and a new acute OI need to be excluded. The diagnostic criteria proposed by French et al. include major and minor criteria. Two major or one major and two minor criteria are required to make the diagnosis.

3.9   Prevention

3.9.1   PMTCT

Prevention of new HIV infections is a global priority, with the elimination of mother to child transmission (EMTCT) a key WHO priority. Several countries have been declared to have eliminated MTCT, including Cuba, Thailand and Armenia. The WHO definition of EMTCT is:

	Minimum requirement
EMTCT indicators	Mother to child transmission of HIV <5% in breastfeeding populations at the end of breastfeeding or <2% in non-breastfeeding populations at the end of 6 weeks
	Annual case rate for pediatric HIV <50/100,000 births

Successful implementation of a PMTCT program with an active strategy to decrease new HIV infections especially in women of childbearing age, planned pregnancies and increasing ART coverage, elimination of mother to child transmission will be a reality.

PMTCT consists of two components, maternal prophylaxis/treatment and infant prophylaxis. The aim is to provide ART throughout the duration of HIV transmission risk to the neonate from conception to the cessation of breastfeeding. Table 3.12 summaries PMTCT strategies from the WHO, US, BHIVA and South African guidelines.

3.9.2 Vaccines

Developing a vaccine against HIV has been an elusive research goal since first identification of the virus. Two main areas of research have emerged: the use of passive immunization (broadly neutralizing antibodies e.g. VCR01) and active immunization. The first vaccine strategy to demonstrate some effect was the RV144 study which demonstrated a vaccine efficacy of 31.2%. The study used a prime-boost approach with initially four doses of ALVAC HIV (vCO1521) followed by two doses of AIDSVAC B/E (gp120) over a 6 month period. Since then several vaccines are in various stages of evaluation as illustrated in Table 3.13.

Table 3.12 PMTCT

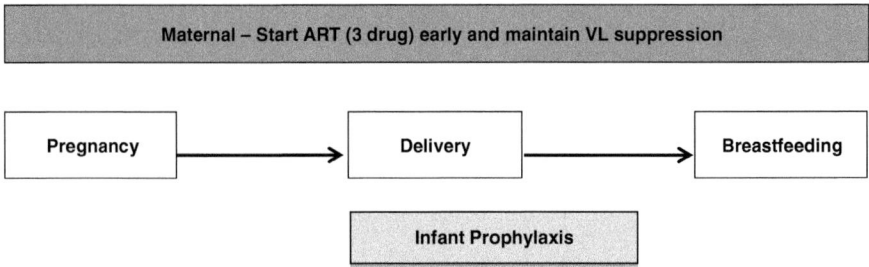

Maternal – Start ART (3 drug) early and maintain VL suppression

Pregnancy	→	Delivery	→	Breastfeeding

Infant Prophylaxis

		WHO[5]	US[17, 19, 20]	PENTA[14, 15]	SA[8]
MATERNAL		ART: Start ALL	ART: Start ALL	ART: Start ALL	ART: Start ALL
		TDF +FTC + EFV	ABC or TDF or AZT + FTC or 3TC + ATV/rtv or DRV/rtv or EFV or RAL	TDF or FTC or ABC + 3TC or EFV + EFV or Boosted PI	TDF + FTC + EFV
INTRA-PARTUM			IV AZT if maternal VL >1000 copies/mL		
INFANT	LOW RISK	NVP or AZT for 4 to 6 weeks	AZT for 4 to 6 weeks	Maternal VL <50 copies/mL AZT for 4 weeks	NVP for 6 weeks
	HIGH RISK	AZT (BD) + NVP for 6 weeks In breastfeeding populations consider for 12 weeks	AZT for 6 weeks NVP (3 doses at birth/48 hrs/96 hrs) 3 drug infant prophylaxis is considered under specialist advice	Maternal VL >50 copies/mL Unbooked mother 3 drug infant prophylaxis for 4 weeks	AZT + NVP for 6 weeks

Table 3.13 HIV vaccine clinical trials snapshot

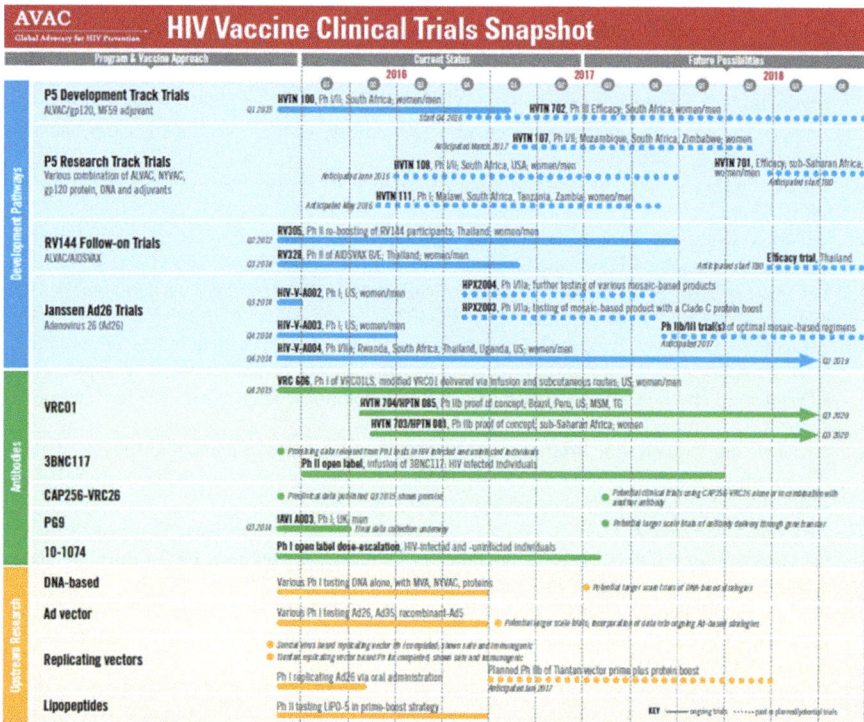

HIV Vaccine Clinical Trials Snapshot

AVAC
Global Advocacy for HIV Prevention

Program & Vaccine Approach	Current Status	Future Possibilities
P5 Development Track Trials — ALVAC/gp120, MF59 adjuvant	HVTN 100, Ph I/II, South Africa; women/men — HVTN 702, Ph IIb Efficacy, South Africa; woman/men	
P5 Research Track Trials — Various combination of ALVAC, NYVAC, gp-120 protein, DNA and adjuvants	HVTN 107, Ph I/II, Mozambique, South Africa, Zimbabwe; women — HVTN 108, Ph I/II, South Africa, USA, women/men — HVTN 111, Ph I, Malawi, South Africa, Tanzania, Zambia, women/men	HVTN 701, Efficacy, sub-Saharan Africa, women/men
RV144 Follow-on Trials — ALVAC/AIDSVAX	RV305, Ph II re-boosting of RV144 participants, Thailand; women/men — RV328, Ph II of AIDSVAX B/E; Thailand; women/men	Efficacy trial, Thailand
Janssen Ad26 Trials — Adenovirus 26 (Ad26)	HIV-V-A002, Ph I; US; women/men — HPX2004, Ph I/IIa; further testing of various mosaic-based products — HPX2003, Ph I/IIa; testing of mosaic-based product with a Clade C protein boost — HIV-V-A003, Ph I; US; women/men — HIV-V-A004, Ph I/IIa; Rwanda, South Africa, Thailand, Uganda, US; women/men	Ph IIb/III trial(s) of optimal mosaic-based regimens
VRC01	VRC 606, Ph I of VRC01LS, modified VRC01 delivered via infusion and subcutaneous routes; US; woman/men — HVTN 704/HPTN 085, Ph IIb proof of concept; Brazil, Peru, US; MSM, TG — HVTN 703/HPTN 081, Ph IIb proof of concept; sub-Saharan Africa; women	
3BNC117	Ph II open label, infusion of 3BNC117; HIV-infected individuals	
CAP256-VRC26		
PG9	IAVI A003, Ph I; UK; men	
10-1074	Ph I open label dose-escalation, HIV-infected and -uninfected individuals	
DNA-based	Various Ph I testing DNA alone, with MVA, NYVAC, proteins	
Ad vector	Various Ph I testing Ad26, Ad35, recombinant Ad5	
Replicating vectors	Ph I replicating Ad26 via oral administration	Planned Ph IIb of Tiantan vector prime plus protein boost
Lipopeptides	Ph II testing LIPO-5 in prime-boost strategy	

References

1. UNAIDS. 4th Annual Report of the Global Plan. 2015 Progress Report on the Global Plan. http://www.unaids.org/en/resources/documents/2015/JC2774_2015ProgressReport_GlobalPlan. Accessed on 21 Sept 2016.
2. UNAIDS. Global AIDS Response Progress Reporting (GARPR). Global AIDS update, 2016. http://www.unaids.org/sites/default/files/media_asset/global-AIDS-update-2016_en.pdf. Accessed 21 Sept 2016.
3. World Health Organisation. Consolidated guidelines on HIV prevention, diagnosis, treatment and care for key populations. 2016 update. http://www.who.int/nutrition/publications/hivaids/9789241597524/en/. Accessed 21 Sept 2016.
4. World Health Organisation. Consolidated guidelines on the use of antiretroviral drugs for treating and preventing HIV infection. http://www.who.int/hiv/pub/arv/arv-2016/en/. Accessed 21 Sept 2016.
5. World Health Organisation. Consolidated guidelines on the use of antiretroviral drugs for treating and preventing HIV infection—recommendations for a Public Health Approach. 2nd ed. 2016. http://www.who.int/hiv/pub/arv/arv-2016/en/. Accessed 23 Sept 2016.
6. World Health Organisation. Elimination of mother-to-child transmission (EMTCT) of HIV and syphilis. June 2016. http://www.wpro.who.int/regional_director/regional_directors_report/2016/media/2016_01_dcd_05_hsi.pdf?ua=1. Accessed 29 Sept 2016.

7. World Health Organisation. Guidelines for an integrated approach to nutritional care of HIV-infected children (6 month-14 years). http://www.who.int/nutrition/publications/hivaids/9789241597524/en/. Accessed 21 Sept 2016.

8. World Health Organisation. Pediatric advocacy. Toolkit: for improved pediatric HIV diagnosis, care and treatment in high HIV prevalence countries and regions. December 2011. http://www.who.int/hiv/pub/pediatric_toolkit2011/en/. Accessed 21 Sept 2016.

9. Goulder P, Lewin S, Leitman E. Paediatric HIV infection: the potential for cure. Nat Rev Immunol. 2016;16(4)259–71.

10. Stein J, Storcksdieck Genannt Bonsmann M, Streeck H. Barriers to HIV cure. HLA. 2016; 88(4)155–63.

11. Violari A, Cotton M, Gibb D, Babiker A, Steyn J, Madhi S, et al. Early antiretroviral therapy and mortality among HIV-infected infants. N Engl J Med. 2008;359(21)2233–44.

12. The Insite START Study Group. Initiation of antiretroviral therapy in early asymptomatic HIV infection. New Engl J Med. 2015;373(9)795–807.

13. B-Lajoie M, Drouin O, Bartlett G, Nguyen Q, Low A, Gavriilidis G, et al. Incidence and prevalence of opportunistic and other infections and the impact of antiretroviral therapy among HIV-infected children in low- and middle-income countries: a systematic review and meta-analysis. Clin Infect Dis. 2016;62(12)1586–94.

14. World Health Organisation. Management of severe malnutrition: a manual for physicians and other senior health workers. 1999. Available at. http://apps.who.int/iris/bitstream/10665/41999/1/a57361.pdf Accessed on 03 March 2017.

15. Bamford A, Turkova A, Lyall H, Foster C, Klein N, Bastiaans D. Paediatric European Network for Treatment of AIDS (PENTA) guidelines for treatment of paediatric HIV-1 infection 2015: optimizing health in preparation for adult life. HIV Med; 2015.

16. De Ruiter A, Taylor GP, Clayden P, Dhar J, Gandhi K, Gilleece Y, et al. British HIV Association guidelines for the management of HIV infection in pregnant women 2012 (2014 interim review)HIV Med. 2014;15:1–77.

17. Department of Health, RSA. HIV guidelines. National consolidated guidelines for the prevention of mother-to-child transmission of HIV (PMTCT) and the management of HIV in children, adolescents and adults. 24 December 2014. http://www.kznhealth.gov.za/family/HIV-Guidelines-Jan2015.pdf. Accessed 21 Sept 2016.

18. Guidelines for the use of antiretroviral agents in pediatric HIV infection. https://aidsinfo.nih.gov/guidelines/html/2/pediatric-arv-guidelines/0. Accessed 23 Sept 2016.

19. Joint United Nations Programme on HIV/AIDS (UNAIDS). 90-90-90 an ambitious treatment target to help end the AIDS epidemic. October 2014. http://www.unaids.org/sites/default/files/media_asset/90-90-90_en_0.pdf. Accessed 29 Sept 2016.

20. Panel on Antiretroviral Therapy and Medical Management of HIV-Infected Children. Guidelines for the use of antiretroviral agents in pediatric HIV infection. https://aidsinfo.nih.gov/contentfiles/lvguidelines/pediatricguidelines.pdf. Accessed 22 Sept 2016.

21. Panel on Treatment of HIV-Infected Pregnant Women and Prevention of Perinatal Transmission. Recommendations for use of antiretroviral drugs in pregnant HIV-1-infected women for maternal health and interventions to reduce perinatal HIV transmission in the United States. https://aidsinfo.nih.gov/contentfiles/lvguidelines/PerinatalGL.pdf. Accessed 22 Sept 2016.

22. Steering Committee PENTA, Welch S, Sharland M, Lyall EG, Tudor-Williams G, Niehues T, Wintergerst U, Bunupuradah T, et al. PENTA 2009 guidelines for the use of antiretroviral therapy in paediatric HIV-1 infection. HIV Med. 2009;10(10)591–613.

Tuberculosis and Other Opportunistic Infections in HIV-Infected Children

4

Helena Rabie and Ben J. Marais

Abstract

Tuberculosis (TB) is the number one infectious disease killer on the planet and children living in human immunodeficiency virus (HIV)-affected households are particularly vulnerable. These children are at increased risk of TB exposure, given the high rates of TB among HIV-infected adults, and those who are very young or have HIV-induced immune compromise are at increased risk of progression to active disease. Apart from TB, HIV-infected children have increased susceptibility to a variety of other infections. Common infections such as viral and bacterial pneumonia occur with increased frequency and severity. They are also prone to opportunistic infections, which refer to organisms with low pathogenic potential that mainly cause disease in people with immune compromise. The main focus of this chapter is on tuberculosis and other opportunistic infections in HIV-infected children.

4.1   Introduction

Children with human immunodeficiency virus (HIV) infection have increased susceptibility to a variety of other infections as a result of HIV-induced immune compromise. Common infections such as viral and bacterial pneumonia occur with

H. Rabie
Department of Paediatrics and Child Health and KIDCRU Pediatric Infectious Diseases
Clinical Research Unit, Stellenbosch University, Tygerberg Children's Hospital,
Cape Town, South Africa

B.J. Marais (✉)
Clinical School, The Children's Hospital at Westmead and the Marie Bashir Institute for
Infectious Diseases and Biosecurity, Locked Bag 4001, Westmead, NSW 2145, Australia

University of Sydney, Sydney, NSW, Australia
e-mail: ben.marais@health.nsw.gov.au

© Springer International Publishing AG 2017
R.J. Green (ed.), *Viral Infections in Children, Volume I*,
DOI 10.1007/978-3-319-54033-7_4

increased frequency and severity. As in HIV-uninfected children, severe respiratory disease is the most common cause of death in HIV-infected children [1, 2]. They are also prone to opportunistic infections, which refer to organisms with low pathogenic potential that mainly cause disease in people with immunodeficiency. Tuberculosis (TB) is often regarded as an opportunistic infection, since disease risk is greatly elevated in those with immunodeficiency [3].

In infants and young children the CD4 T-lymphocyte count is poorly predictive of HIV disease progression and death [4, 5]. Therefore all infants and young children require antiretroviral therapy (ART) to prevent HIV disease progression, irrespective of absolute CD4 count, CD4 percentage or clinical staging (Table 4.1). It is important to note the frequency and severity of infections, before and after, ART initiation, including seemingly common infections such as otitis media and fungal

Table 4.1 World Health Organization clinical HIV staging

Stage 1	Asymptomatic; generalised lymphadenopathy
Stage 2	Unexplained persistent hepatosplenomegaly
	Papular pruritic eruptions; fungal nail infection
	Angular cheilitis; lineal gingival erythema; recurrent oral ulcerations
	Extensive wart virus infection or molluscum contagiosum
	Unexplained persistent parotid enlargement
	Herpes zoster
	Recurrent or chronic upper respiratory tract infections; otitis media, otorrhea, sinusitis or tonsillitis
Stage 3	Unexplained moderate malnutrition or wasting
	Unexplained persistent diarrhea (14 days or more)
	Unexplained persistent fever (intermittent or constant, for longer than 1 month)
	Persistent oral candidiasis (after first 6–8 weeks of life)
	Oral hairy leukoplakia; acute necrotizing ulcerative gingivitis or periodontitis
	Lymph node or pulmonary tuberculosis
	Severe recurrent bacterial pneumonia or chronic lung disease
	Symptomatic lymphoid interstitial pneumonitis
	Unexplained anemia, neutropenia and/or chronic thrombocytopenia
Stage 4	Unexplained severe wasting, stunting or severe malnutrition
	Pneumocystis pneumonia
	Recurrent severe bacterial infections
	Chronic herpes simplex infection
	Esophageal or airway/lung candidiasis
	Extrapulmonary tuberculosis
	Kaposi's sarcoma
	Cytomegalovirus infection with onset at age older than 1 month
	Central nervous system toxoplasmosis (after 1 month of life)
	Extrapulmonary cryptococcosis (including meningitis)
	HIV encephalopathy
	Disseminated endemic mycosis or non-tuberculous mycobacterial infection

Table 4.1 (continued)

Chronic cryptosporidiosis or isosporiasis (with diarrhea)
Cerebral or B-cell non-Hodgkin's lymphoma
Progressive multifocal leukoencephalopathy
Symptomatic HIV-associated nephropathy or HIV-associated cardiomyopathy
HIV-associated rectovaginal fistula (African children)
Reactivation of American trypanosomiasis (South American children)

skin infections, to help monitor clinical progress and assess therapeutic efficacy [5]. Although effective ART leads to consistent and major reductions in infections, there is often a transient increase in the incidence and/or severity of disease during the first 2–3 months, especially in those with poor immune function. This is referred to as the immune reconstitution inflammatory syndrome (IRIS) [6].

In tuberculosis endemic settings, tuberculosis and bacterial pneumonia remain the most common infections in children even after ART initiation [7–10]. The increased risk of bacterial pneumonia and sepsis that persists despite ART, highlights the importance of optimal vaccination, and where possible revaccination, after ART initiation and immune recovery [10]. Non-life threatening infections can also be troublesome i.e., severe *Molluscum contagiosum* on the face can be disfiguring and stigmatizing and recurrent otitis media can cause hearing loss. Among HIV-infected children adopted to families in the United States, *M. contagiosum*, fungal infections of the skin and intestinal parasites were regarded as the most troublesome infectious diseases [11]. Table 4.2 provides an overview of common infections in HIV-infected children, although diseases with geographically defined incidence, such as malaria, should also be considered. This chapter focuses mainly on TB and other common opportunistic infections.

4.2   Tuberculosis

Globally, the TB epidemic is sustained in conditions of poverty, but it has been fueled by the immune compromise resulting from HIV infection, particularly in sub-Saharan Africa. The epicenter of TB/HIV co-infection is in Southern Africa (Fig. 4.1), where most TB patients are HIV co-infected and TB incidence rates regularly exceed 500/100,000 population (Fig. 4.2) [12]. Children contribute little to the maintenance of the TB or HIV epidemics, but they are greatly affected by it. It is estimated that one million children developed TB in 2014, with the majority of cases occurring in Africa, South-East Asia and the Western Pacific [12]. In HIV-endemic areas woman of child bearing age are disproportionately affected by HIV and TB [13, 14], implying a high risk of household TB exposure for vulnerable young children [15]. In an isoniazid preventative therapy (IPT) trial in South Africa, 10% of HIV-exposed infants had contact with a potential TB source case by 14 weeks of age [16], and a TB incidence of 1.6% (1596 cases per 100,000) has been reported in HIV-infected infants, in the absence of universal early ART [17]. If an HIV-infected women develops TB during pregnancy the infant is at risk of TB if the

Table 4.2 Overview of microorganisms and clinical disease syndromes observed in HIV-infected children[a]

Organism	Clinical syndrome
Bacteria	
Haemophilus influenzae	U/LRTI, meningitis, bacteremia
Streptococcus pneumoniae	U/LRTI, meningitis, bacteremia
Klebsiella pneumoniae	U/LRTI, UTI, meningitis, bacteremia
Salmonella spp.	LRTI, gastroentritis, meningitis, bacteremia
Escherichia coli	LRTI, UTI, meningitis, bacteremia
Staphylococcus aureus	LRTI, meningitis, bacteremia, bone skin and joint
Viruses	
Common respiratory viruses	U/LRTI
Cytomegalovirus (CMV)	Congenital infection, pneumonia, retinitis, esophagitis, colitis
Herpes simplex	Skin and soft tissue, ulcers, keratitis, encephalitis
EBV	Cancers
Varicella zoster	Chickenpox and zoster, keratitis, retinitis, encephalitis
Influenza	U/LRTI
Molluscum contagiosum	Skin infections
Human papiloma virus	Skin, mouth and upper airway infections, genital warts, cancer risk
JC and BK virus	Progressive multifocal leucoencephalopathy (rarely reported in children)
Measles	Increased risk of severe disease
Mycobacteria	
Mycobacterium tuberculosis	All forms of tuberculosis
M. bovis BCG	Young infants not on ART are at risk of BCG dissemination and children starting ART of BCG IRIS
M. avium-intracellulare (MAC)	Chronic lung infection
Fungi	
Cryptococcus spp.	CNS, lung, skin infection
Candida spp.	Mucosal, nail and skin infection
Pneumocystis jiroveci (PCP)	Chronic lung infection
Protozoa	
Crypto and microsporidium	Gastrointestinal and bile duct infection
Toxoplasma gondii	Congenital infection and CNS disease

HIV human immunodeficiency virus, *URTI* upper respiratory tract infection, *LRTI* lower respiratory tract infection, *UTI* urinary tract infection, *PLE* progressive multifocal leucoencephalopathy, *ART* antiretroviral therapy, *BCG* Bacille Calmette-Guerin, *IRIS* immune reconstitution inflammatory syndrome, *CNS* central nervous system
[a]Especially in children with poor disease control and significant immune compromise

mother is inadequately treated, in addition the infant also experiences a threefold higher risk of vertical HIV infection [18]. In the absence of strategies to actively identify TB in HIV-infected pregnant women there may be significant delays in TB diagnosis with greatly increased risk to the mother and her baby [14].

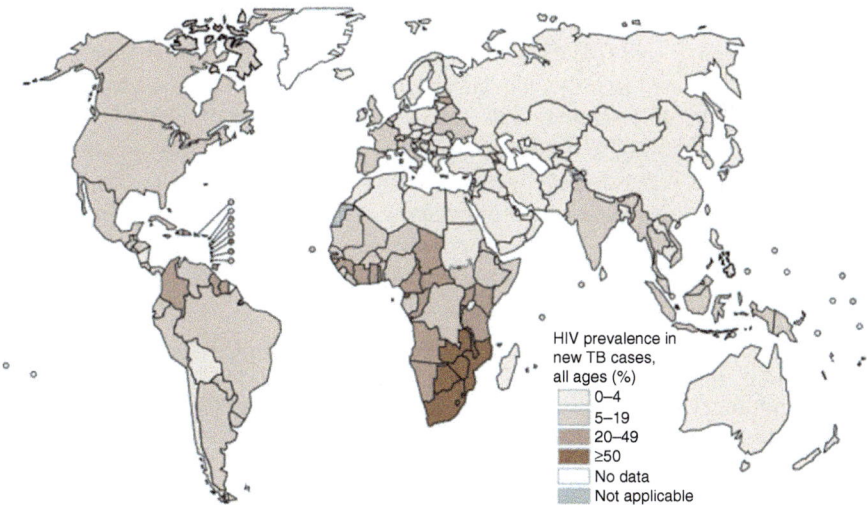

Fig. 4.1 Estimated HIV prevalence in tuberculosis patients (2014) (From the World Health Organization Global Tuberculosis Report 2015 [12]). *HIV* human immunodeficiency virus

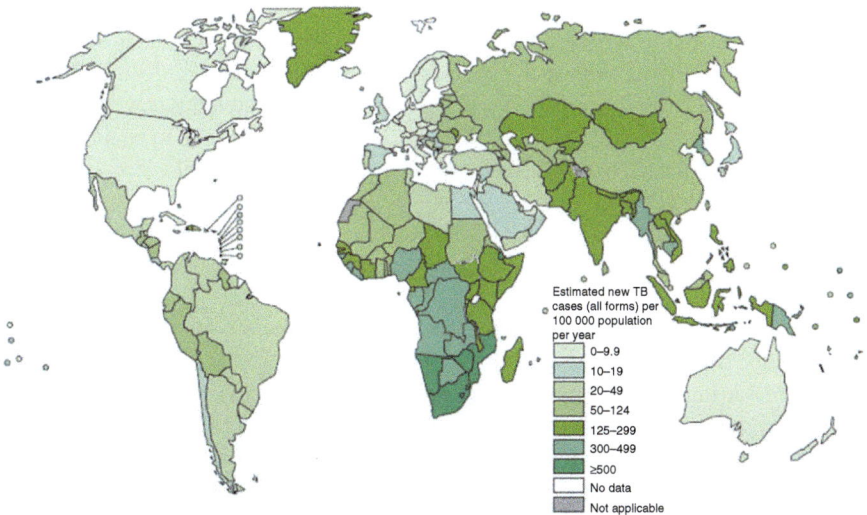

Fig. 4.2 Global tuberculosis incidence estimates (2014) (From the World Health Organization Global Tuberculosis Report 2015 [12])

4.2.1 Prevention

Immunization is usually the most effective from of primary prevention. *M. bovis* bacille Calmette-Guerin (BCG) vaccination is routinely given to newborn babies in TB endemic settings, with at-birth vaccine delivery reaching more than 80% of children [19]. Although BCG provides significant protection against disseminated forms of TB in early childhood, the level of protection is variable, with no evidence of protection in immunocompromised children [20]. Since BCG is an attenuated live vaccine, HIV-infected children are at increased risk of developing local or disseminated BCG disease [21]. Simian models also indicate that BCG vaccination provides a strong T-cell stimulus that may increase HIV viral replication [22]. However, given the major benefits attributed to BCG in TB endemic settings, current recommendations are to continue at-birth vaccination of all children, while ensuring optimal prevention of mother-to-child transmission of HIV (PMTCT) and remaining vigilant of potential BCG-associated complications [23].

The most important tools available to reduce the TB risk in HIV-infected children are (1) early and universal access to ART for all children with HIV and (2) use of IPT in children with documented TB infection or exposure. A randomized controlled trial of early or delayed ART in young infants, the Children with HIV Early Antiretroviral (CHER) trial, identified TB in 8% of children receiving early ART and in 20% of children with delayed ART during the first year of life [4]. ART initiation in older children is also associated with major reductions in incident TB, although TB disease risk remains elevated compared to uninfected peers [24]. Improved ART access for adults also benefits children, as demonstrated by concomitant declines in adult and paediatric TB caseloads with increased ART uptake. From 2005 to 2009, adult ART access increased from 21.5 to 68.2% in Soweto, South Africa, and during the same time TB in HIV-uninfected children declined from 18.7 to 11.0 per 100,000 population [25]. The ambitious new "90-90-90" targets defined by UNAIDS aims for 90% of HIV-infected persons to be diagnosed, 90% of those diagnosed to be on ART and 90% of those on ART to have an undetectable HIV viral load by 2020 [26].

IPT following documented TB exposure is a well-established TB prevention strategy [27]. There are no randomized controlled trials to assess the benefit of post-exposure IPT in HIV-infected children, although this can be extrapolated from studies in HIV-uninfected children and HIV-infected adults. A study that assessed the value of universal IPT in HIV-infected children living in a high TB incidence setting, documented major reductions in all-cause mortality (16% vs. 8%) and incident TB (9.9% vs. 3.8%) compared to placebo [28]. Additional analysis demonstrated that the greatest reductions occurred in children who used both ART and IPT, establishing the additive value of IPT [29]. A subsequent multicentre trial that assessed the value of universal IPT in very young infants found no difference in disease rates between IPT and placebo, which complicates the guidance in HIV-infected infants without documented TB exposure [30]. The Antiretroviral Research for Watoto (ARROW) study in Zimbabwe followed 1206 HIV-infected children for 5 years after ART initiation and included randomized discontinuation of co-trimoxazole. TB risk was significantly associated with the duration of adherent ART, but surprisingly also with co-trimoxazole use, which remains an intriguing observation [31].

4.2.2 Diagnosis

The diagnosis of TB in HIV-infected children remains challenging, due to the pauci-bacillary nature of childhood tuberculosis and the overlap with other HIV-related diseases [32]. Recent (in the past 12 months) contact with an infectious TB case (including those with sputum smear-negative pulmonary TB) is essential to explore. Clinical criteria for the diagnosis of TB, such as prolonged coughing, failure to thrive, or fatigue are less helpful in HIV-infected children in whom these symptoms are common and therefore less specific [33]. In addition, TB-associated symptoms are often more acute in HIV-infected children, reducing the sensitivity of chronic symptoms [34]. However, non-remitting symptoms or a change in baseline symptomatology, especially if accompanied by unexplained weight loss, should always prompt the clinician to consider a diagnosis of active TB. On examination, physical signs are often non-specific. As in HIV-uninfected children, the most frequent intra-thoracic disease manifestation is hilar and/or paratracheal adenopathy [35]. In general, the clinical spectrum of disease recorded in HIV-infected children with immune compromise, is similar to that seen in very young (<3 years of age) children with immune immaturity, with increased frequency of advanced and/or disseminated (miliary) disease [36]. Table 4.3 summarizes the different clinical syndromes associated with TB infection and disease.

Due to poor sensitivity, a tuberculin skin test (TST) has reduced value to detect TB infection in HIV-infected children. Severe malnutrition, low CD4 counts and progressive HIV disease are associated with false negative TST results. Novel interferon gamma release assays (IGRAs) are more specific than the TST, but sensitivity is broadly similar [37]. Table 4.4 compares key characteristics of the TST and commercially available IGRAs. As with the TST a positive IGRA result confirms TB infection, which may aid the diagnosis of active TB or indicate the need for preventive therapy, but fails to make the crucial distinction between latent infection and active disease. In addition, neither test has adequate sensitivity to serve as a "rule-out" test for TB disease. It may be difficult to distinguish TB from other HIV-associated lung conditions on CXR. Mediastinal and/or paratracheal lymphadenopathy may occur with lymphocytic interstitial pneumonitis (LIP) and other mycobacterial or fungal infections. The typical picture of disseminated (miliary) TB may also be mimicked by the reticulonodular pattern of LIP, although this rarely extends into the peripheral lung fields. Disease in a very young child (<2 years) and an evenly spread, fine, nodular pattern on CXR favours a diagnosis of disseminated (miliary) TB, while clubbing, generalized lymphadenopathy and swollen parotids in an older child suggests LIP (Table 4.5).

Bacteriologic cultures or nucleic acid amplification testing (NAAT) are the only ways to establish a definitive diagnosis [37, 38]. Respiratory specimens should ideally include at least two fasting, early morning gastric aspirates or induced sputum samples. Sampling of all potential sources should be considered, including fine needle aspiration biopsy (FNAB) of peripheral masses, pleural or cerebrospinal

Table 4.3 Clinical syndromes associated with tuberculosis in children

Pathological classification	Disease phase (time period)	Clinical syndromes	Risk groups[a]	Pathogenesis	Imaging manifestations
Primary *M. tb* infection	Incubation (0–6 weeks)	Asymptomatic	All ages	No adaptive immunity TST(−); IGRA(−)	None
	Infection (1–3 months)	Self-limiting symptoms (mild, viral-like)		Adaptive immunity IGRA(+); TST(+)	Transient hilar or mediastinal lymphadenopathy (50–70% of cases), rarely visible transient Ghon focus
		Hypersensitivity reactions (fever; erythema nodosum; phlyctenular conjunctivitis)		No test to register reinfection	
Early disease progression >90% of disease occur within 12 months of primary infection	Very early (2–6 months)	Uncomplicated lymph node (LN) disease	<10 years	Inadequate innate and/or adaptive immunity	Hilar or mediastinal lymphadenopathy without airway or parenchymal involvement
		Progressive Ghon focus	<1 year	TST(+); IGRA(+)	Ghon focus with visible cavitation
		Disseminated disease:	<3 years	May be negative with immune compromise or extensive disease, cannot be used as "rule out" tests	
		–Miliary disease			–Discrete lung nodules (1–2 mm) on CXR; hepato-splenomegaly
		–TB Meningitis			–Hydrocephalus; basal enhancement; brain infarcts and/or tuberculomas
	Early (4–12 months)	Complicated LN disease	<5 years		–Hyperinflation or atelectasis/collapse
		–Airway compression			
		–Expansile caseating pneumonia			–Expansile consolidation of a segment or lobe
		–Infiltration of adjacent anatomic structures (esophagus, phrenic nerve, pericardium)			–Tracheo-/broncho-esophageal fistula; pericardial effusion; hemidiaphragmatic palsy
		Pleural disease	>3 years		Effusion usually unilateral; some pleural thickening and loculations (due to fibrinous strands)
		–Exudative effusion (rarely empyema; or chylothorax)			
		Lymphadenitis	1–10 years		Usually not needed, matting and edema of adjacent soft tissue
		–Most common extra-thoracic manifestation			

Late disease progression Generally rare apart from adult-type disease in adolescents	Late (1–3 years)	Adult-type pulmonary disease −Difficult to differentiate primary infection; reactivation and reinfection disease	≥8 years	"Overaggressive" innate and/or adaptive immunity	Apical cavities; may be bilateral; minimal or no LN enlargement (previously referred to as post-primary TB)
		Osteoarticular disease: −Spondylitis/arthritis/osteomyelitis	≥5 years	Inadequate local control; usually local manifestations only, but can disseminate from any active focus	Periarticular osteopenia, subchondral cystic erosions, joint space narrowing
	Very late (>3 years)	Urinary ract (renal, ureter, bladder) disease	>5 years		Renal calcifications; hydronephrosis, calyceal dilation and/or ureter stricture

Age ranges, risk groups and timelines provide general guidance only

aHIV-infected children with immunocompromise are particularly vulnerable, similar to children <3 years of age, and may present with atypical features.

Adapted from [36]

Table 4.4 Comparison of T-cell based tests for *M. tuberculosis* infection

Characteristic	Tuberculin skin test	QuantiFERON-TB®	T-SPOT-TB®
Complexity	Low	Moderate	High
TB antigens	PPD-tuberculin (not *M. tb* specific)	ESAT-6; CFP-10; TB-7.7	ESAT-6; CFP-10
Method/measurement	Intradermal injection skin induration after *in vivo* stimulation	Blood draw	Blood draw
		ELISA-based measurement of IFN-γ production by T-cells after *in vitro* stimulation	ELISPOT-based measurement of IFN-γ-producing T-cells (spots) after *in vitro* stimulation
Number of visits	2 visits (within 48–72 h)	1 visit	
Operator dependent	Yes	No	
Cross reactivity			
BCG	Yes[a]	No	
NTM	Yes	Rare (*M. kansasii, M. marinum, M. szulgaii*)	
Boosting with repeat testing	Yes	No	
Internal controls	No	Yes	
Utility by age	Reduced sensitivity <6 months of age	Reduced sensitivity <5 years of age	
Sensitivity[b] (HIV uninfected)	Around 80%	Around 80%	
Sensitivity[b] (HIV-infected)	Less than 50%	Less than 50%; T-SPOT.TB slightly less affected by immunosuppression	

BCG Bacille Calmette-Guérin, *IGRA* interferon-gamma release assay, *M. tb Mycobacterium tuberculosis*, *NTM* nontuberculous mycobacteria, *PPD* purified protein derivative, *TB* tuberculosis, *TST* tuberculin skin test, adapted from [37]
[a](Particularly if vaccinated after infancy or repeatedly)
[b](Cannot be used as a "rule-out" test)

fluid, ear swabs or bone marrow aspirates etc. [37]. Table 4.6 provides a brief overview of benefits and limitations associated with various specimen collection methods. Although TB treatment is frequently instituted on clinical suspicion, it remains important to perform initial cultures, including repeat cultures if the response to treatment is poor, to confirm the diagnosis where possible and screen for possible drug resistant TB.

4.2.3 Treatment

The standard TB treatment regimen consists of isoniazid (INH), rifampicin (RMP), pyrazinamide and ethambutol during the 2-month intensive phase followed by INH

Table 4.5 Features that may help to differentiate *Pneumocystis jiroveci* pneumonia (PJP), bacterial pneumonia and pulmonary tuberculosis in HIV-infected children[a]

Feature	PJP	*Bacterial pneumonia*	TB
Age	Infants (mostly <6 months)	All ages	All ages (mostly >6 months)
Fever	Low-grade	High-grade	Intermittent
Hypoxia	Severe and persistent	Variable and acute	Uncommon
Rapidity of onset	Days	Days	Weeks but can present acutely
Response to antibiotics	Poor if not initially considered and treated	Good but there are concerns	Poor
Auscultation	Variable	Signs of focal consolidation	Variable
Chest radiograph	Usually diffuse interstitial	Usually focal	Usually intra-thoracic lymphadenopathy
Response to appropriate treatment	Slow / Outcome in general poor	Fast / Outcome in general good	Slow / Outcome in general good

PJP Pneumocystis jiroveci pneumonia, *TB* tuberculosis
[a]Adapted from [32]; these features only offer general guidance and atypical presentations are common

Table 4.6 Tuberculosis specimen collection methods—perceived problems and benefits[a]

Specimen collection method	Problems/benefits	Potential clinical application
Sputum	Not feasible in very young children; assistance and supervision may improve the quality of the specimen	Routine sample to be collected in children >7 years of age (all children who can produce a good quality specimen)
Induced sputum	Comparable yield to gastric aspirate; no age restriction; specialized technique, which requires nebulization and suction facilities; potential transmission risk	To be considered in the hospital setting on an in- or out-patient basis
Gastric aspirate	Unpleasant procedure, but not difficult to perform; requires fasting / Sample collection advised on 3 consecutive days	Routine sample to be collected in hospitalized who cannot produce a good quality sputum specimen
Nasopharyngeal aspiration	Less invasive than gastric aspirate; no fasting required; comparable yield to gastric aspirate	To be considered in primary health care clinics or on an outpatient basis
String test	Less invasive than gastric aspirate; tolerated well in children >4 years; bacteriologic yield and feasibility requires further investigation	Potential to become the routine sample collected in children who can swallow the capsule, but cannot produce a good quality sputum specimen

(continued)

Table 4.6 (continued)

Specimen collection method	Problems/benefits	Potential clinical application
Broncho-alveolar lavage	Extremely invasive	Only for use in patients who are intubated or who require diagnostic bronchoscopy
Stool	Culture not practical, DNA extraction difficult	Reasonable yield using Gene Xpert®
	Not invasive; *M. tuberculosis* excretion well documented	
Urine	Not invasive; excretion of *M. tuberculosis* components	Lipoarabinomanan (LAM) assay has poor sensitivity; unreliable in children
Blood/bone marrow	Good sample sources to consider in the case of probable disseminated TB	To be considered for the confirmation of probable disseminated TB in hospitalized patients
Cerebrospinal fluid (CSF)	Fairly invasive; bacteriologic yield low	To be considered if signs of tuberculous meningitis
Fine needle aspiration biopsy (FNAB)	Minimally invasive using a fine 23G needle; excellent bacteriologic yield, minimal side-effects	Procedure of choice in children with superficial lymphadenopathy

[a]Adapted from [38]

and RMP during the 4-month continuation phase. Due to well-documented cases of TB relapse in HIV-infected children, prolonging the treatment duration to 9 months may have benefit, especially in children with significant immune compromise [39]. TB recurrence (relapse or reinfection) is not uncommon and if response to treatment is poor or a second episode of TB is clinically suspected, every effort should be made to confirm the diagnosis on culture and to do drug susceptibility testing (DST). The emergence of drug resistant TB and on-going transmission of these strains within communities pose a grave risk to global TB control and to vulnerable HIV-infected children [40]. Drug resistant TB should be considered if the child shows poor TB treatment response, or had contact with someone who died on TB treatment, had poor treatment response or documented drug resistant TB [41]. In general children with drug-resistant TB achieve better outcomes than adults [42], but treatment is complicated and should be discussed with an expert.

4.2.3.1 TB Treatment and ART

ART is an essential component of TB case management in HIV-infected children. In those not already on ART, initiation should be considered as soon as possible; usually after 2 weeks of TB treatment in children who are not severely immune compromised. Therapeutic failure should be considered in children on ART who develop incident TB. All rifamycins, including rifampicin, induce P450 enzyme systems in the liver and intestinal wall, resulting in increased elimination of protease inhibitors (ritonavir is least affected and partially reverses the effect), and non-nucleoside reverse transcriptase inhibitors, particularly nevirapine. Rifampicin also promotes glucuronidation of zidovudine and probably abacavir, but its clinical significance is uncertain. A number of strategies could be employed to manage the drug-drug interactions between TB treatment and ART [43]. Two nucleoside reverse transcriptase inhibitors (NRTIs) and efavirenz can be continued without adjustment, and is the preferred ART regimen during TB co-treatment. If nevirapine is the only non-NRTI available, dose adjustment should be considered for the duration of rifampicin treatment. Children on lopinavir/ritonavir should receive additional rito-navir. Although it is not an optimally suppressive regimen, the use of three NRTIs (zidovudine, lamivudine and abacavir) may offer a short-term option if drug-drug interactions is a problem. Table 4.7 summarizes the use of first and second-line TB drugs in children on ART. Although adverse effects related to TB drugs and ART may be similar, severe adverse effects are fortunately uncommon. Most adverse

Table 4.7 Use of first and second-line TB treatment with antiretrovirals in children with TB/HIV co-infection

TB drugs	Recommended dose	Drug-drug interactions with ART
First-line treatment		
Isoniazid	7–15 mg/kg once daily; max 300 mg	None
Rifampicin	10–20 mg/kg once daily; max 600 mg	Reduces plasma levels of NNRTIs, PIs and integrase inhibitors
Pyrazinamide	30–40 mg/kg once daily; max 2 g daily	None
Ethambutol	15–25 mg/kg once daily	None
Rifabutin	10–20 mg/kg/day; max 300 mg	Boosted PI: increase rifabutin levels; NNRTI: Efavirence reduces the rifabutin levels; no dose adjustment with nevirapine
Second-line treatment		
Injectables		
Kana/capreomycin	15–30 mg/kg once daily; max 1 g	Unlikely
Amikacin	15–22.5 mg/kg once daily; max 1 g	Unlikely
Streptomycin	20–40 mg/kg once daily; max 1 g	Should not be used in children
Flouroquinolones		
Levofloxacin	15–20 mg/kg once daily; max 750 mg	Buffered didanosine may reduce oral absorption of all fluoroquinolones
Moxifloxacin	7.5–10 mg/kg once daily; max 400 mg	Ritonavir may reduce moxifloxacin levels

(continued)

Table 4.7 (continued)

TB drugs	Recommended dose	Drug-drug interactions with ART
Other drugs with reasonable potency[a]		
Pro/ethionamide	15–20 mg/kg once daily; max 1 g	Possible
Cycloserine/terizidone	10–20 mg/kg once daily, max 1 g	Unlikely
Linezolid	10 mg/kg 1–2 times daily	Unlikely
Other drugs with uncertain/poor potency		
Clofazimine	3–5 mg/kg once daily	Unknown; may be a weak CYP3A4 inhibitor
Meropenem/clavulanic acid Imipenem/cilastin	As for bacterial infections	Unlikely
Thiacetazone	5–8 mg/kg once daily	Contraindicated in HIV-infected individuals
Para-aminosalicylic acid (PAS)	150–200 mg/kg granules daily in two divided doses, max 12 g	Efavirenz may reduce PAS levels

TB tuberculosis, *ART* antiretroviral therapy, *NNRTI* non-nucleoside reverse transcriptase inhibitor, *PI* protease inhibitor, *AUC* area under the curve, *HIV* human immunodeficiency virus
[a]Newer drugs such a bedaquiline and delamanid are not included, since more information is required on potential drug-drug interactions. Adapted from [43]

effects occur within 2 months of starting TB therapy, including hepatotoxicity, skin rashes, nausea and vomiting, leukopenia and/or anemia, and peripheral neuropathy. Pyridoxine supplementation is advised for all HIV-infected children on TB treatment and ART to reduce the risk of peripheral neuropathy.

4.2.3.2 Immune Reconstitution Inflammatory Syndrome (IRIS)

IRIS describes the temporary exacerbation of TB symptoms and signs that may occur when ART is first introduced, or the "unmasking" of disease in someone with undiagnosed tuberculosis [44]. It usually manifests within 3 months of ART initiation and subsides spontaneously [45]. IRIS presents like a "disease" exacerbation, but it is not an indication of treatment failure. IRIS of the central nervous system is a particular concern and it may be prudent to delay the onset of ART in HIV-infected cases with with tuberculous meningitis (TBM) [46].

4.3 Other Opportunistic Infections

4.3.1 *Pneumocystis Jiroveci* Pneumonia (PJP)

Pneumocystis pneumonia is frequently the first serious illness encountered by the HIV-infected infant not on co-trimoxazole prophylaxis [47]. Autopsy studies indicate that PJP is the leading cause of death in HIV-infected children <6 months of age; declining in importance after 12–18 months of age [1]. Co-trimoxazole prophylaxis protects against severe disease rather than infection, as *P. jiroveci* has been isolated from 38% of children considered to have received adequate prophylaxis [48]. Typical presenting features include tachypnoea, mild intermittent fever and a dry cough with minimal auscultatory findings [49]. The radiographic picture may be highly variable [50], but significant hypoxia is typical and the lactate dehydrogenase (LDH) level is usually markedly elevated, although this is not specific. In high HIV burden settings all infants requiring hospitalization for pneumonia require review of their HIV status. The treatment of choice is high-dose oral or intravenous co-trimoxazole (20 mg/kg/day of the trimethoprim component, divided into four doses) for a total duration of 2–3 weeks. If only oral therapy is available an oral loading dose of 20 mg/kg stat is recommended [51]. The addition of corticosteroids (prednisone 2 mg/kg/day for 1–2 weeks and then tapered) and oxygen supplementation is advised in all children with hypoxia (saturation levels <92% in room air). Response to therapy is relatively slow, but co-trimoxazole resistance seems rare.

4.3.2 Molluscum Contagiosum

Disease caused by *Molluscum contagiosum* commonly manifests as self-limiting dome shaped papules with a characteristic central umbilication, which takes 12–18 months to disappear. In HIV-infected children skin involvement can be extensive, with giant nodular tumours or epidermal cysts, and include atypical sites such as the neck and face. Scratching of lesions causes autoinoculation and should be strongly discouraged. Keratoconjunctivitis can develop secondary to lesions on the eyelids and it may also physically obscure the child's vision; in these cases an ophthalmologist should be consulted [52, 53]. Therapeutic options include topical salicylic or trichloroacetic acid, curettage, cautery and cryotherapy. Referral to a dermatologist may be helpful with extensive lesions, but no treatment is truly effective [54]. Established lesions may initially increase in size following ART initiation, but this is due to IRIS and ART should not be discontinued [52, 53].

4.3.3 Human Papilloma Virus

Skin warts are the most common manifestation of human papilloma virus (HPV) infection in children with HIV; lesions may be wide spread and often involve the mucosal surfaces. The presence of genital warts raises concern of sexual abuse if diagnosed in young pre-adolescent children, but may also results from perinatal exposure, autoinoculation and other forms of contact spread [55]. Warts in the mouth may appear as multiple, smooth raised lumps resembling focal epithelial hyperplasia [56]. Occasionally warts may cause upper airway obstruction; this condition is very difficult to treat and should be referred for expert management. Epidermodysplasia verruciformis lesions have also been described [52]. Effective ART is the most important treatment modality; local treatment options include topical application of 25% podophyllin, 80% trichloroacetic acid or imiqiod cream, or tissue destruction with liquid nitrogen, cautery or laser. One of the most important concerns with HPV infection is the potential for malignant transformation observed with some serotypes, especially those causing genital and oral lesions. Fortunately, available HPV vaccines elicit adequate immune responses in HIV-infected adolescents and young adults on ART [57, 58].

4.3.4   Herpes Simplex Virus

Young children acquire herpes simplex virus type-1 (HSV-1) from oral secretions of caretakers or playmates with herpes stomatitis or "fever blisters", although asymptomatic HSV-1 shedding also occurs in infected individuals. Primary HSV-1 infection commonly results in extensive, but self-limiting (within 5–7 days), gingivostomatitis in immune competent children. HIV-infected children experience similar extensive gingivostomatitis, but it may take a very long time to heal. Ulcers that persist for more than 4 weeks is an AIDS defining condition, while painful recurrences are not uncommon and may occur as part of IRIS [52]. Treatment with acyclovir may shorten the course of illness and in children with frequent recurrences it is advisable to provide them with oral valacyclovir to initiate treatment as soon as symptoms appear. Disseminated infection with HSV-1 can cause encephalitis and/or acute hepatic failure. One should suspect dissemination in any acutely ill child with herpetic lesions in the mouth and features of meningoencephalitis or acute hepatic dysfunction. The cerebrospinal fluid (CSF) in herpes encephalitis shows a lymphocytic pleocytosis with slightly raised protein. A positive polymerase chain reaction (PCR) test for HSV-1 in the CSF is the definitive test. Therapy is most effective when started early and consists of intravenous (IV) acyclovir 20 mg/kg/dose 8 hourly for 21 days. HSV 1-associated keratitis, conjuctivitis and retinitis may also occur and should be aggressively managed to prevent permanent eye damage [10].

4.3.5   Varicella Zoster Virus (VZV)

Chickenpox and zoster are different disease manifestations caused by the same virus. Vaccination to prevent varicella is recommended in children with HIV who have a CD4 of >15%. Prior ART initiation increases the efficacy of the vaccine [10]. Although active zoster lesions are contagious they are much less so than chickenpox lesions. Chickenpox is highly infectious and also associated with aerosol transmission. Children with chickenpox or zoster that require hospitalization should be isolated until all the vesicles have crusted, this may take longer in immune compromised children. If susceptible children have a history of exposure with in the past 3–4 days prevention strategies must be considered. It is important to note that prevention may fail and that the onset of symptoms may be delayed for up to 28 days in cases where preventative measures were instituted.

In children with HIV, chickenpox manifestations are often more severe and more frequently complicated by secondary bacterial infection. Those with significant immunocompromise tend to have atypical rash with new crops of lesions presenting for weeks or months, often with hemorrhagic vesicles and involvement of the palms and soles. Lesions can develop into non-healing ulcers that become necrotic, crusted, and hyperkeratotic. Dissemination to the central nervous system, lungs and liver may occur [59–61]. Treatment with high dose oral acyclovir may reduce the period of infectivity and shorten the course of illness, although IV acyclovir is recommended for severely ill children. VZV is more resistant to acyclovir than HSV-1 and therefore, the recommended standard treatment dose is 20 mg/kg/dose 8 hourly [10].

Zoster represents a "flare-up" of the chickenpox virus that has remained "dormant" within the anterior horn cell of the sensory nerve. In immune competent persons it is usually a self-limited disease involving a single dermatome in older people, but HIV-infected children may experience multiple relapses or develop chronic lesions that involve more than one dermatome. With severe immune compromise, encephalitis and retinitis could occur as part of the zoster relapse [59–61]. The condition is best treated with high dose oral acyclovir and adequate pain relief. If carbamazepine is deemed necessary to help control chronic pain then the usual dose escalation regimen should be followed to prevent adverse drug reactions and drug-drug interactions with ART should be taken into consideration. Active vaccination will prevent infection and may also be used as post exposure prevention. Passive vaccination with varicella antibodies (VZIG) is indicated for children who lack immunity and are severely immune suppressed, if administered within 3–5 days of exposure. Antiviral prophylaxis with acyclovir offers an alternative beyond 7–10 days of exposure, but has not been studied in HIV-infected children.

4.3.6 Cytomegalovirus

Cytomegalovirus (CMV) infection is ubiquitous with high levels of viral shedding in the genital tracts of HIV-infected women. HIV exposed infants may present with congenital CMV or become infected at a young age. A positive CMV culture or PCR from nasopharyngeal aspirates and/or urine indicates infection and viral shedding, but may also be found in children who are completely asymptomatic. The diagnosis of actual target organ CMV disease is more problematic, depending on relevant signs and symptoms together with high levels of CMV viremia, as indicated by quantitative DNA PCR test results. Histopathology remains the most definitive diagnostic test, but it is unfeasible in most settings. CMV can affect virtually any organ system, but pneumonia, retinitis and gastrointestinal disease are most common. CMV and PCP co-infection can make it difficult to ascertain the contribution of CMV to the underlying lung pathology in children with severe symptoms. Due to the risk of immune reconstitution disease, the optimal timing of instituting ART in infants with CMV pneumonia remains a topic of debate, but should be considered as soon as disease symptoms have subsided on treatment. The treatment of CMV consists of intravenous ganciclovir or oral valganciclovir. Ganciclovir may cause significant bone marrow suppression and patients should be regularly assessed while receiving high doses. The cost and complexity of treatment limits access in low resource settings.

CMV retinitis should be suspected in a severely immune compromised child that report changes in visual acuity and/or "floaters". In small infants retinal involvement may be indicated by an acquired inability to fix and follow and/or absent light reflexes. Intra-ocular injections with ganciclovir or foscarnet implants have been used with great success in adults, with some evidence of success in children where it is technically more difficult. CMV infection may also cause hepatitis, cholangitis, or cholecystitis. Jaundice may be evident and liver enzymes are often increased. CMV may involve any part of the intestine, but the colon is most frequently involved. CMV colitis often presents with unexplained chronic diarrhea, abdominal pain and fever. The stools often contain blood (frank or occult) and the abdominal film may show a dilated colon; diagnosis is confirmed by colonoscopy and biopsy. Colon perforation and strictures are rare complications.

4.3.7 Cryptococcosis

Most cases of cryptococcosis in HIV-infected patients are caused by *Cryptococcus neoformans*. Infection is acquired through inhalation and these fungi have a high affinity for the central nervous system [62]. For reasons that are unknown, children have a lower incidence of cryptococcal meningitis than adults, with a bimodal distribution in infants (<12 months) and children aged 5–10 years of age [63]. Because pediatric disease is rare, therapy recommendations are extrapolated from adult studies.

Cryptococcal meningitis is the most common disease manifestation and should always be considered in children with severe immune depletion. Neurological disease usually has an indolent nature with fever, headache and vomiting, and no neck stiffness, but more acute presentations with seizures and focal neurological signs, including neck stiffness has been reported [62]. Lumbar puncture opening pressure is usually elevated but the cerebrospinal fluid (CSF) shows few changes. CSF India ink wet mounts have largely been replaced by a cryptococcal antigen tests that have better sensitivity [62]. Cryptococcal antigen screening tests are recommended in all adults with severe immune suppression and symptoms requiring CSF collection; in children it should be considered [64]. Treatment consists of amphotericin, flucytocine and fluconazole, followed by secondary fluconazole prevention. High dose fluconazole treatment is an option in settings where amphotericin and flucytocine are unavailable [64]. It is important to manage the raised intracranial pressure though repeated lumbar puncture. The use of corticosteroids has been associated with adverse outcomes and is not recommended during initial therapy [65], although observational data suggests that it may be of benefit later in the treatment course [66]. Children with cryptococcal meningitis who are not yet on ART should not start ART for at least 3 weeks after the initiation of effective antifungal treatment. Adult data demonstrated high rates of CNS IRIS with serious complications if ART is started earlier [67]. Pulmonary and other forms of disseminated disease may occur, in which case a lumbar puncture should be performed to exclude concurrent CNS infection.

4.3.8 Tinea, Onychomycosis and Candida

Fungal infections of the skin nails and hair, as well as, mucosal surfaces are common in HIV-infected children and occur despite ART [10]. In a series of children adopted to the US from low income settings, 39% had tinea capitis despite the fact that the majority were on ART at the time of adoption [11]. Superficial skin hair and nail infections can be recalcitrant to therapy. Diagnosis requires culture and histopathological examination of skin, or hair and nail clippings. Typically onychomycosis affects fingernails and is accompanied by paronychia. Systemic azole therapy for weeks to months (depending on the site) is often needed. If these medications are prescribed, drug-drug interactions with ART should be reviewed [10].

Oral/pharyngeal candidiasis is common in HIV-infected infants, irrespective of early ART [7]. However, esophageal candidiasis is rarely seen when a child is on ART. Esophageal candidiasis presents with dysphagia, retrosternal pain, nausea, and vomiting; in young babies it may present with pooling of milk when swallowing and posturing with feeds. Tracheobronchial candidiasis can present with hoarseness and upper airways obstruction, sometimes in the absence of oral/pharyngeal candida, while systemic candida is usually only seen children with central venous catheters. Treatment should consider the severity of disease and the area affected. Systemic antifungal therapy is required for all infections, except superficial skin or mouth involvement [10].

4.3.9 Cryptosporidium

Cryptosporidium hominis and *parvum* commonly cause self-limiting diarrhea. However, with advanced immune suppression it may cause protracted diarrhea that can be life threatening. It may also cause bile duct inflammation and obstructive jaundice; diagnosis is through the detection of oocysts in the stool. Supportive care and treatment with ART is important, since immune recovery is essential to clear the infection.

4.3.10 Progressive Multifocal Leukoencephalopathy

Progressive multifocal leukoencephalopathy (PML) is a fatal, subacute, demyelinating disease of the central nervous system caused by a reactivation of the JC polyomavirus, which induces a lytic infection of oligodendrocytes. Fortunately it rarely affects children. The onset of PML is insidious, but the disease leads to death in 3–6 months if ART is not initiated. In the early phases there is impairment of speech and vision and mental deterioration, but in advanced stages paralysis of limbs, cortical blindness, and sensory abnormalities are common. CSF investigations are often normal, but the virus may be detected by PCR. Magnetic resonance imaging (MRI) is the diagnostic modality of choice and shows white matter lesions without swelling, mass effect or contrast enhancement. ART is the only effective therapeutic option, although some disease manifestations may be IRIS-related [68].

4.3.11 Kaposi Sarcoma

Kaposi sarcoma, multicentric Castleman disease (non-malignant tumors of the lymphnodes) and primary effusion lymphomas (B-cell lymphoma affecting body cavities) may result as a consequence of human herpes virus 8 (HHV-8) infection [69, 70]. Like Epstein Barr virus (EBV), HHV-8 is a gamma herpes virus that may increase the proliferation of infected cells. Transmission is mainly through saliva and infection is common in many Mediterranean and African countries. In Southern Africa up to 50% of HIV-infected persons are co-infected with HHV-8, placing them at risk for Kaposi sarcoma [71–73]. Kaposi sarcoma is the most commonly reported cancer in HIV-infected children (34 per 100,000 person-years in South African) [74, 75]. The classic disease presentation is with blue/purple papules on the skin or oral mucosa, but children frequently present with mass lesions in the upper airways leading to airway obstruction, recurrent episodes of gastrointestinal or pulmonary bleeding, or persistent pleural effusion [76, 77]. The definitive management is immune reconstitution with ART and chemotherapy is only considered in severe cases or with worsening symptoms on ART; a pediatric oncologist should be consulted [10, 78].

4.4   Conclusion

HIV-infected children with severe immune compromise are vulnerable to many opportunistic infections and rare complications, however, it is important to remember that common things occur commonly. This chapter has highlighted some of the common and more severe conditions that are seen in this group to raise clinical awareness and improve case management. The most important interventions should focus on the prevention of HIV infection in infants and young children, ART initiation at the earliest possible time point in those that are infected and ongoing treatment adherence to ensure successful viral suppression.

References

1. Bates M, Mudenda V, Mwaba P, Zumla A. Deaths due to respiratory tract infections in Africa: a review of autopsy studies. Curr Opin Pulm Med. 2013;19:229–37.
2. Gupta RK, Lucas SB, Fielding KL, Lawn SD. Prevalence of tuberculosis in post-mortem studies of HIV infected adults and children in resource-limited settings: a systematic review and meta-analysis. AIDS. 2015;29:1987–2002.

3. Venturini E, Turkova A, Chiappini E, Galli L, de Martino M, Thorne C. Tuberculosis and HIV co-infection in children. BMC Infect Dis. 2014;14(Suppl 1):S5.
4. Violari A, Cotton MF, Gibb DM, et al. Early antiretroviral therapy and mortality among HIV infected infants. N Engl J Med. 2008;359:2233–44.
5. World Health Organization. Consolidated guidelines on the use of antiretroviral drugs for treating and preventing HIV infection: recommendations for a public health approach. Geneva: World Health Organization; 2016. http://apps.who.int/iris/handle/10665/208825
6. Boulware DR, Callens S, Pahwa S. Pediatric HIV immune reconstitution inflammatory syndrome. Curr Opin HIV AIDS. 2008;3:461–7.
7. B-Lajoie MR, Drouin O, Bartlett G, et al. Incidence and prevalence of opportunistic and other infections and the impact of antiretroviral therapy among HIV-infected children in low- and middle-income countries: a systematic review and meta-analysis. Clin Infect Dis. 2016;62:1586–94.
8. Nesheim SR, Kapogiannis BG, Soe MM, et al. Trends in opportunistic infections in the pre- and post-highly active antiretroviral therapy eras among HIV-infected children in the Perinatal AIDS Collaborative Transmission Study, 1986-2004. Pediatrics. 2007;120:100–9.
9. Dankner WM, Lindsey JC, Levin MJ, Pediatric ACTGPT. Correlates of opportunistic infections in children infected with the human immunodeficiency virus managed before highly active antiretroviral therapy. Pediatr Infect Dis J. 2001;20:40–8.
10. Panel on Opportunistic Infections in HIV-Exposed and HIV-Infected Children. Guidelines for the prevention and treatment of opportunistic infections in HIV-exposed and HIV-infected children. Department of Health and Human Services. http://aidsinfo.nih.gov/contentfiles/lvguidelines/oi_guidelines_pediatrics.pdf
11. Wolf ER, Beste S, Barr E, Wallace J, et al. Health outcomes of international HIV-infected adoptees in the United States. Pediatr Infect Dis J. 2016;35:422–7.
12. World Health Organization. Global tuberculosis report 2015. WHO/HTM/TB/2015.
13. Marais BJ, Gupta A, Starke JR, El Sony A. Tuberculosis in women and children. Lancet. 2010;375:2057–9.
14. Lawn SD, Bekker LG, Middelkoop K, Myer L, Wood R. Impact of HIV infection on the epidemiology of tuberculosis in a peri-urban community in South Africa: the need for age-specific interventions. Clin Infect Dis. 2006;42:1040–7.
15. Wood R, Johnstone-Robertson S, Uys P, et al. Tuberculosis transmission to young children in a South African community: modeling household and community infection risks. Clin Infect Dis. 2010;51:401–8.
16. Cotton MF, Schaaf HS, Lottering G, Weber HL, Coetzee J, Nachman S. Tuberculosis exposure in HIV-exposed infants in a high-prevalence setting. Int J Tuberc Lung Dis. 2008;12:225–7.
17. Hesseling AC, Cotton MF, Jennings T, et al. High incidence of tuberculosis among HIV-infected infants: evidence from a South African population-based study highlights the need for improved tuberculosis control strategies. Clin Infect Dis. 2009;48:10.
18. Gupta A, Bhosale R, Kinikar A, et al. Maternal tuberculosis: a risk factor for mother-to-child transmission of human immunodeficiency virus. J Infect Dis. 2011;203:358–63.
19. Marais BJ, Seddon JA, Detjen AK, et al. Interrupted BCG vaccination is a major threat to global child health. Lancet Respir Med. 2016;4:251–3.
20. Fordham von Reyn C. Routine childhood Bacille Calmette Guerin immunization and HIV infection. Clin Infect Dis. 2006;42:559–61.
21. Hesseling AC, Rabie H, Marais BJ, et al. Bacille Calmette-Guerin (BCG) vaccine-induced complications and HIV infection in children. Clin Infect Dis. 2006;42:548–58.
22. Hesseling AC, Cotton MF, Marais BJ, et al. BCG and HIV reconsidered: moving the research agenda forward. Vaccine. 2007;25:6565–8.
23. Hessling AC, Cotton MF, Fordham von Reyn C, Graham SM, Gie RP, Hussey GD. Consensus statement on the revised World Health Organization recommendations for BCG vaccination in HIV-infected infants. Int J Tuberc Lung Dis. 2008;12:1376–9.
24. Walters E, Cotton MF, Rabie H, Schaaf HS, Walters LO, Marais BJ. Clinical presentation and outcome of tuberculosis in human immunodeficiency virus infected children on antiretroviral therapy. BMC Pediatr. 2008;8:1.

25. Dangor Z, Izu A, Hillier K, et al. Impact of the antiretroviral treatment program on the burden of hospitalization for culture-confirmed tuberculosis in South African Children: a time-series analysis. Pediatr Infect Dis J. 2013;32:972–7.

26. UNAIDS. 90–90–90 An ambitious treatment target to help end the AIDS epidemic. UNAIDS/ JC2684; 2014.

27. Marais BJ, Ayles H, Graham SM, Godfrey-Faussett P. Screening and preventive therapy for tuberculosis. Clin Chest Med. 2009;30:827–46.

28. Zar HJ, Cotton MF, Strauss S, et al. Effect of isoniazid prophylaxis on mortality and incidence of tuberculosis in children with HIV: randomised controlled trial. BMJ. 2007;334:136.

29. Frigati LJ, Kranzer K, Cotton MF, et al. The impact of isoniazid preventive therapy and anti-retroviral therapy on tuberculosis in children infected with HIV in a high tuberculosis inci-dence setting. Thorax. 2011;66:496–501.

30. Madhi SA, Nachman S, Violari A, et al. Effect of primary isoniazid prophylaxis against tuber-culosis in HIV-exposed children. N Engl J Med. 2011;365:21–31.

31. Crook AM, Turkova A, Musiime V, et al. Tuberculosis incidence is high in HIV infected African children but is reduced by co-trimoxazole and time on antiretroviral therapy. BMC Med. 2016;14:50.

32. Marais BJ, Graham SM, Cotton MF, Beyers N. Diagnostic and management challenges of childhood TB in the era of HIV. J Infect Dis. 2007;196(Suppl 1):S76–85.

33. Marais BJ, Gie RP, Schaaf HS, et al. A refined symptom-based approach to diagnose pulmo-nary tuberculosis in children. Pediatrics. 2006:e1350–9.

34. Moore DP, Klugman KP, Madhi SA. Role of Streptococcus pneumoniae in hospitalization for acute community-acquired pneumonia associated with culture-confirmed Mycobacterium tuberculosis in children: a pneumococcal conjugate vaccine probe study. Pediatr Infect Dis J. 2010;29:1099–104.

35. Marais BJ, Gie RP, Schaaf HS, et al. The natural history of childhood intra-thoracic tuberculo-sis: a critical review of literature from the pre-chemotherapy era. Int J Tuberc Lung Dis. 2004;8:392–402.

36. Perez-Velez CM, Marais BJ. Tuberculosis in children. N Engl J Med. 2012;367:348–61.

37. Roya-Pabon CL, Perez-Velez CM. Tuberculosis exposure, infection and disease in children: a systematic diagnostic approach. BMC Pneumonia 2016;8:23.

38. Marais BJ, Pai M. New approaches and emerging technologies in the diagnosis of childhood tuberculosis. Paediatr Respir Rev. 2007;8:124–33.

39. Schaaf HS, Krook S, Hollemans DW, et al. Recurrent culture-confirmed tuberculosis in human immunodeficiency-virus infected children. Pediatr Infect Dis. 2005;24:685–91.

40. Marais B. The global tuberculosis situation and the inexorable rise of drug-resistant disease. Adv Drug Deliv Rev. 2016;102:3–9.

41. Schaaf HS, Marais BJ. Management of multidrug-resistant tuberculosis in children: a survival guide for paediatricians. Paediatr Respir Rev. 2011;12(3):1–8.

42. Isaakidis P, Casas EC, Das M, Tseretopoulou X, Ntzani EE, Ford N. Treatment outcomes for HIV and MDR-TB co-infected adults and children: systematic review and meta-analysis. Int J Tuberc Lung Dis. 2015;19:969–78.

43. Rabie H, Gousaard P. Tuberculosis and pneumonia in HIV infected children, an overview. Pneumonia. 2016;8:19.

44. Meintjes G, Rabie H, Wilkinson RJ, Cotton MF. Tuberculosis-associated immune reconstitu-tion inflammatory syndrome and unmasking of tuberculosis by antiretroviral therapy. Clin Chest Med. 2009;30:797–810.

45. Link-Gelles R, Moultrie H, Sawry S, Murdoch D, Van Rie A. Tuberculosis immune reconstitu-tion inflammatory syndrome in children initiating antiretroviral therapy for HIV infection: a systematic literature review. Pediatr Infect Dis J. 2014;33:499–503.

46. van Toorn R, Rabie H, Dramowski A, Schoeman JF. Neurological manifestations of TB-IRIS: a report of 4 children. Eur J Paediatr Neurol. 2012;16:676–82.

47. Chintu C, Bhat GJ, Walker AS, et al. Co-trimoxazole as prophylaxis against opportunistic infections in HIV infected Zambian children (CHAP): a double-blind randomised placebo-controlled trial. Lancet. 2004;364:1865–71.

48. Morrow BM, Samuel CM, Zampoli M, Whitelaw A, Zar HJ. Pneumocystis pneumonia in South African children diagnosed by molecular methods. BMC Res Notes. 2014;7:26.
49. Graham SM, Mtitimila EI, Kamanga HS, Walsh AL, Hart CA, Molyneux ME. Clinical presentation and outcome of Pneumocystis carinii pneumonia in Malawian children. Lancet 2000; 355:369–73.
50. Pitcher RD, Daya R, Beningfield SJ, Zar HJ. Chest radiographic presenting features and radiographic progression of pneumocystis pneumonia in South African children. Pediatr Pulmonol. 2011;46:1015–22.
51. Zar HJ, Langdon G, Apolles P, et al. Oral trimethoprim-sulphamethoxazole levels in stable HIV-infected children. S Afr Med J. 2006;96:627–9.
52. Mankahla A, Mosam A. Common skin conditions in children with HIV/AIDS. Am J Clin Dermatol. 2012;13:153–66.
53. Guidelines on the Treatment of Skin and Oral HIV-Associated Conditions in Children and Adults. Geneva: World Health Organization; 2014.
54. Van der Wouden JC, van der Sande R, van Suijlekom-Smit LWA, et al. Interventions for cutaneous molluscum contagiosum. Cochrane Database Syst Rev. 2009;7(4):CD004767.
55. Jones V, Smith SJ, Omar HA. Nonsexual transmission of anogenital warts in children: a retrospective analysis. Scientific World Journal. 2007;7:1896–9.
56. Moscicki AB, Farhat S, Yao TJ, Ryder MI, et al. Oral human papillomavirus in youth from the pediatric HIV/AIDS cohort study. Sex Transm Dis. 2016;43:498–500.
57. Levin MJ, Moscicki AB, Song LY, et al. Safety and immunogenicity of a quadrivalent human papillomavirus (types 6, 11,16, and 18) vaccine in HIV-infected children 7 to 12 years old. J Acquir Immune Defic Syndr. 2010;55(2):197–204.
58. Weinberg A, Song LY, Saah A, et al. Humoral, mucosal and cell-mediated immunity against vaccine and non-vaccine genotypes after administration of quadrivalent human papillomavirus vaccine to HIV-infected children. J Infect Dis. 2012;206:1309–18.
59. Jura E, Chadwick EG, Josephs SH, et al. Varicella-zoster virus infections in children infected with human immunodeficiency virus. Pediatr Infect Dis J. 1989;8:586–90.
60. Srugo I, Israele V, Wittek AE, et al. Clinical manifestations of varicella-zoster virus infections in human immunodeficiency virus-infected children. Am J Dis Child. 1993;147:742–5.
61. Wood SM, Shah SS, Steenhoff AP, Rutstein RM. Primary varicella and herpes zoster among HIV-infected children from 1989 to 2006. Pediatrics. 2008;121:e150–6.
62. Kao C, Goldman DL. Cryptococcal disease in HIV-infected children. Curr Infect Dis Rep. 2016;18:27.
63. Meiring ST, Quan VC, Cohen C, et al. A comparison of pediatric- and adult-onset cryptococcosis detected through population surveillance in South Africa, 2005–2007. AIDS. 2012;26:2307–14.
64. WHO. Rapid advice: diagnosis, prevention and management of cryptococcal disease in HIV-infected adults, adolescents and children. 2011. http://www.who.int/hiv/pub/cryptococcal_disease2011/en/
65. Beardsley J, Wolbers M, Kibengo FM, Ggayi AB, et al. Adjunctive dexamethasone in HIV-associated cryptococcal meningitis. N Engl J Med. 2016;374:542–54.
66. Beardsley J, Wolbers M, Day JN, CrptoDex Investigators. Dexamethasone in cryptococcal meningitis. N Engl J Med. 2016;375:189–90.
67. Guidelines for the Prevention and Treatment of Opportunistic Infections in HIV-Exposed and HIV-Infected Children: Panel on Opportunistic Infections in HIV-Exposed and HIV-Infected Children. Department of Health and Human Services. http://aidsinfo.nih.gov/contentfiles/lvguidelines/oi_guidelines_pediatrics.pdf
68. Schwenk H, Ramirez-Avila L, Sheu SH, et al. Progressive multifocal leukoencephalopathy in pediatric patients: case report and literature review. Pediatr Infect Dis J. 2014;33(4):e99–105.
69. Chang Y, Cesarman E, Pessin MS, Lee F, Culpepper J, Knowles DM, Moore PS. Identification of herpesvirus-like DNA sequences in AIDS-associated Kaposi's sarcoma. Science. 1994;266:1865.
70. Orem J, Otieno MW, Remick SC. AIDS-associated cancer in developing nations. Curr Opin Oncol. 2004;16:468.

71. Gantt S, Orem J, Krantz EM, Morrow RA, et al. Prospective characterization of the risk factors for transmission and symptoms of primary human herpesvirus infections among Ugandan infants. J Infect Dis. 2016;214:36.
72. Butler LM, Were WA, Balinandi S, Downing R, et al. Human herpesvirus 8 infection in children and adults in a population-based study in rural Uganda. J Infect Dis. 2011;203:625.
73. Rohner E, Wyss N, Heg Z, et al. HIV and human herpesvirus 8 co-infection across the globe: systematic review and meta-analysis. Int J Cancer. 2016;138:45–54.
74. Bohlius J, Maxwell N, Spoerri A, Wainwright R, et al. Incidence of AIDS-defining and other cancers in HIV-positive children in South Africa: record linkage study. Pediatr Infect Dis J. 2016;35:e164–70.
75. Rohner E, Valeri F, Maskew M, Prozesky H, et al. Incidence rate of Kaposi sarcoma in HIV-infected patients on antiretroviral therapy in Southern Africa: a prospective multicohort study. J Acquir Immune Defic Syndr. 2014;67:547–54.
76. Luppi M, Barozzi P, Rasini V, et al. Severe pancytopenia and hemophagocytosis after HHV-8 primary infection in a renal transplant patient successfully treated with foscarnet. Transplantation. 2002;74:131–2.
77. Vaz P, Macassa E, Jani I, et al. Treatment of Kaposi sarcoma in human immunodeficiency virus-1-infected Mozambican children with antiretroviral drugs and chemotherapy. Pediatr Infect Dis J. 2011;30:891–3.
78. World Health Organization. Guidelines on the treatment of skin and oral HIV-associated conditions in children and adults. Geneva; 2014.

Emerging Zoonotic and Vector-Borne Viral Diseases

5

Jacqueline Weyer and Lucille H. Blumberg

Abstract

Many vector-borne and zoonotic diseases are considered to be emerging; since they are either newly reported to cause human disease, or are causing disease in geographical locations or species not previously documented. In the past 15 years, significant outbreaks of Severe Acute Respiratory Syndrome (or SARS) and Middle Eastern Respiratory Syndrome (or MERS), Nipah and Hendra, Ebola virus disease and Zika fever and others have been reported. In this chapter the clinical characteristics, epidemiological aspects, treatment and prevention and information related to the laboratory investigation of important zoonotic and vector-borne diseases that have emerged in the past 10 years, and how this affects children, will be discussed. Furthermore rabies, considered a neglected viral disease with the majority of victims in Africa being children, will also be addressed.

5.1 Hemorrhagic Fevers

The viral hemorrhagic fevers (VHF) are a specific group of diseases that have in common the propensity for human to human transmission particularly in the hospital setting, have high mortality and require early recognition and a public health response to prevent outbreaks. They are caused by unrelated RNA viruses belonging to the *Filovirus, Arenavirus, Bunyavirus* and *Flavivirus* genera (reviewed in [1, 2]) (Table 5.1). Subclinical or mild cases are commonly associated with for example Lassa fever, yellow fever and Crimean-Congo hemorrhagic fever (CCHF), contrasted by Ebola virus and Marburg virus disease (EVD and MVD) which are more often

J. Weyer (✉) • L.H. Blumberg
National Institute for Communicable Diseases, Johannesburg, South Africa
e-mail: jacquelinew@nicd.ac.za; lucilleb@nicd.ac.za

© Springer International Publishing AG 2017
R.J. Green (ed.), *Viral Infections in Children, Volume I*,
DOI 10.1007/978-3-319-54033-7_5

Table 5.1 Summary of medically important and most commonly occurring viral hemorrhagic fevers

Viral hemorrhagic fever	Crimean-Congo hemorrhagic fever	Marburg virus disease	Lassa fever	Yellow fever
Reported geographic distribution	Most wide spread tick-borne viral infection in humans. Reported from more than 50 countries in Africa, South East Europe, Middle East and Asia. Including China, Turkey, Bulgaria, Croatia, Greece, Iran, Iraq, Saudi Arabia, Kenya, Uganda, Tanzania, Egypt, Nigeria and South Africa	Angola, Democratic Republic of Congo, Kenya, Uganda, Zimbabwe	Sierra Leone, Guinea, Liberia and Nigeria	Endemic in more than 50 countries in tropical regions of Africa (44 countries) and South and Central America (12 countries)
Cumulative occurrence/ prevalence	Sero-prevalence of up to 20% in endemic areas in people with history of tick bites	Only 466 cases reported from 1967 to 2014 (including cases associated with exportation of infected monkeys to Europe, Russia and USA). Outbreaks often involve single cases. Two bigger outbreaks reported in DRC and Angola in 1998–2000 and 2004–2005, respectively	Up to 500,000 cases in West Africa per annum. Sero-prevalence of up to 55% in certain communities in endemic areas	Up to 130,000 cases with hemorrhagic manifestation per year by estimation
Mode of transmission	Mostly through tick bites, or other contact with infected ticks, or infected tissues and blood. Nosocomial transmission has been reported, but human-to-human transmission limited	Spill over from cave dwelling bats, exact mode of transmission unclear. Human-to-human transmission through direct and close contact with infected persons	Human contact with infected rodent urine and fecal material. The latter may be aerosolized and transmitted *via* this route	From day feeding *Aedes* mosquitoes, particularly *A. aegypti*

Reservoir host and vector	Domestic livestock and small mammals Ticks, particularly *Hyalomma* ticks	Believed to be cave dwelling bats such as *Roussettus aegyptiacus*	*Mastomys natalensis* (or the multimammate mouse)	Maintained in sylvatic cycle in nature involving certain primates and canopy dwelling mosquitoes. May establish in urban setting involving *Aedes aegypti*
Case fatality rate	Up to 30% in hospitalized cases	Up to 90% in outbreaks	Up to 20% in hospitalized cases, but up to 80% of people infected do not develop signs and symptoms of Lassa fever	Up to 50% in hospitalized cases
References	Uyar et al. [13], Ergönül [14], Bente et al. [15], Mardani and Pourkaveh [16], Elaldi and Kaya [17], Messina et al. [18], Shayan et al. [19], Leblebicioglu et al. [20]	Brauburger et al. [21], Olival and Hayman [22], Brainard et al. [23], Rougeron et al. [24]	Richmond and Baglole [25], Ogbu et al. [26]	Gardner and Ryman [27], Garske et al. [28]

Ebola virus disease is not included in this summary, but described in detail in the text. The geographical expanse provided for each disease related to reports of naturally occurring outbreaks and does for example not include exported cases of disease associated with severe and fatal outcomes and case fatality rates of up to 90% recorded during outbreaks.

All of the HF viruses are naturally harbored in distinct animal or invertebrate reservoirs. Outbreaks of the disease are either associated with exposure to the virus through a vector species (for example ticks for CCHF virus or mosquitoes for yellow fever) and little or no human-to-human spread, or a single spill-over event from the animal reservoir with amplification of the outbreak through human-to-human transmission (for example spill-over of Ebola virus from bats or other infected forest dwelling animals to humans). Since these viruses are associated with an animal or invertebrate reservoirs and vectors, the diseases have a specifically geographical expanse, partly related to the distribution of these reservoir and vector species. Another observation is that historically, outbreaks of VHFs have been occurring in low resource settings with many of the outbreaks associated with health care facilities where poor infection control practices allow for the spread of these viruses. Recent outbreaks of yellow fever have occurred in high density urban areas with intense mosquito breeding. Infection prevention and control measures in hospitals remain the most important measure to contain a VHF outbreak [3–7]. This is in the absence of prophylaxis or treatment on the account of most of the VHFs. Currently, the only VHF for which a licensed vaccine is available is yellow fever. The vaccine is administered to children over 9 months of age of as part of the expanded program of immunization in a number of yellow fever endemic countries, in mass campaigns and in response to reported outbreaks. Documentation of a high rate of yellow fever vaccine-associated neurotropic disease in young infants during the 1960s led to institution of age <6 months as a contraindication for yellow fever vaccination except during epidemics when the risk of yellow fever virus transmission may be very high [8].

Ribavirin may be useful for treating Lassa fever and CCHF, although not proven in clinical trials [9–12].

A number of VHF infections such as CCHF are related to occupational exposure to infected vectors or animals so infections in children are less common.

For the purpose of this chapter further discussions will focus on aspects of Ebola virus disease (EVD).

5.1.1 Epidemiological Aspects of EVD

To date, natural outbreaks of EVD have only been recorded in central and west African countries. Before the outbreak of EVD in Guinea, Sierra Leone and Liberia from 2013 to 2015, outbreaks of the disease were limited to secluded and rural locations in central Africa [29]. A cumulative 31,139 cases with 12,890 deaths have been recorded since the first recognition of EVD in 1976–2015, more than 90% of these cases were reported during the West Africa outbreak [30, 31]. Single exported cases during outbreaks have been recorded in travellers with secondary transmission noted during four such events in South Africa, Nigeria, USA and Mali [29]. The common perception is that children are spared during EVD outbreaks, and in fact this observation was made during several outbreak investigations [32]. This does however not mean that children are not affected. A retrospective study revealed that 20 of the 218 (9%) confirmed EVD cases during the 2000 Gulu, Uganda outbreak were in young children and adolescents [33]. Nearly 20% of the cases reported during the West Africa outbreak involved children below the age of 15 [34]. The putative index case of the West Africa outbreak was a 2 year old child that may have had exposure to bats whilst playing in the forest nearby his home [35, 36]. The risk of EVD in children may be related to cultural factors in handling and caring for ill persons within households. In communities where children are kept away from the sick the risk should be lower, but in many instances young children are co-admitted to hospital with their sick caregivers increasing the risk of them also being infected [33]. During the Gulu-outbreak, it was noted that the under-five age group was more affected during this outbreak than other children. Here children were typically co-admitted to hospital with their symptomatic parents [33].

The natural history of Ebola virus (EBOV) remains an enigma despite intensive research on the topic. Evidence is accumulating that certain forest dwelling fruit bat species (including *Hypsignathus monstrosus*, *Epomops franqueti*, *Myonycteris torquata* and possibly *Eidolon helvum*) are natural reservoirs of the virus although this occurrence is still not fully understood [37–39]. The source of EVD outbreaks has often been associated with direct contact with bushmeat [40]. The latter includes bats, chimpanzee, gorillas and duiker antelope [41]. The risk for contact with bushmeat appears to be related to slaughtering and preparation of the raw meat and not necessarily the consumption of cooked meat [41]. Outbreaks of EVD in the human population are amplified by human-to-human transmission through close and direct contact with infected individuals [23]. These outbreaks are perpetuated in the hospital setting but also in close family and friend circles where caring for the ill and traditional burial practices are the most prominent risk factor for contracting the virus.

5.1.2 Clinical Aspects of EVD

The initial stage of the clinical course of EVD is non-specific, complicating early diagnosis, and follows an incubation period of 2–21 days [42]. The difficulty of clinical diagnosis is compounded by a long list of possible differential diagnoses in the locations where EVD has been described. This includes other VHF such as Lassa fever, but also other infectious diseases such as malaria, cholera and typhoid [42–45]. EVD is characterized by sudden onset of fever (>40 °C (104 °F)), generalized weakness and anorexia, myalgia, headache and sore throat [42, 46, 47]. These were all common findings in children during the EVD outbreak in West Africa [34]. Noteworthy, is that all of the EVD cases reported in children during the 2000 Gulu outbreak reported fever [33]. Patients infected with EVD rapidly deteriorate and commonly present with gastrointestinal symptoms including vomiting, diarrhea and dehydration. Chertow et al. [48] reported profuse watery diarrhea in EVD patients persisting for a week or more. As with other VHF, and contrary to the naming of the syndrome, most cases do not present with overt bleeding [33, 34, 42, 48]. Less than 20% of children diagnosed with EVD presented with bleeding which was reported as hematochezia, hematemesis, bleeding gums, epistaxis and oozing from needle puncture sites [33, 34]. Patients will then either recover or deteriorate, often displaying neurological signs. Patients typically succumb due to multi-organ failure and hypovolemic shock [42]. Common findings in EVD patients with a poor prognosis are rapidly developing leukopenia, thrombocytopenia and a decrease in coagulopathy markers [42, 49, 50].

The case fatality rate in children is high [51]. During the Gulu-outbreak, the reported case fatality rate among children and adolescents was 40% [33]. During the West Africa outbreak, the case fatality rate in the under-five group was significantly higher than other age groups, with the highest death rate reported in children under-1 year old (90%) [34].

Pregnant woman are more likely to develop severe EVD, and miscarriage is reported universally in pregnant woman diagnosed with EVD [48, 52]. A case report of EVD in a pregnant EVD woman that survived, but still aborted suggests that maternal immunity may have little effect on affording protection to the unborn [53].

5.1.3 Laboratory Diagnosis of EVD

Limited capacity for specialized laboratory diagnosis exists in most countries where EVD has been previously reported. This remains one of the biggest challenges in effective outbreak responses for the future [54, 55]. Laboratory investigations are typically carried out by national reference centres and involve specialized testing to detect the virus, or the body's response to the infection [24, 54, 56]. Specific laboratory confirmation is by reverse transcription (RT)-PCR which detects the viral RNA in a patient's sample. Antigen detection using enzyme immunosorbent assays (ELISA) are also commonly used. Live virus may also be isolated in cell culture, but typically serves as a complimentary assay and due to the time required for results, not always useful for diagnosis or a timely outbreak response. Detection of specific IgM responses or rise in specific IgG titre is also indicative of recent infection. The latter is tested using indirect immunofluorescence assays or ELISA. Some laboratories may offer additional testing such as electron microscopy and immunohistochemistry on tissue samples, but this may be viewed as complimentary testing in most cases. It is recommended that blood and tissues of suspected EVD cases be handled in biosafety level 4 facilities; although RT-PCR testing during the West Africa EVD outbreak was done under field conditions working under modified biosafety level 3 conditions. The West Africa outbreak has highlighted the need for rapid and sensitive tests for EVD. Tests that can be conducted at facility level or at the bedside will be helpful in directing clinical decisions [55, 57]. Although testing at reference laboratories may be laborious and more time consuming than facility-level testing, it is still recommended that all investigations be confirmed by experienced reference laboratories. Several rapid EVD assays are being developed in response to the West Africa outbreak, but are not yet widely available [58–63].

5.1.4 Treatment and Prevention of EVD

Despite tremendous efforts during and following the West Africa outbreak to develop antiviral treatment and vaccines, neither are (widely) available yet [64–66]. Management remains supportive in most instances, although experimental treatment options have been applied for some [67, 68]. Fever and pain is managed with paracetamol and it is recommended to avoid nonsteroidal antipyretics [12]. Chertow et al. [48] who managed more than 700 EVD patients in Liberia in 2014 found that the use of antiemetics, antidiarrheal medications, and rehydration therapy was beneficial. Antimalarial treatment is recommended for all suspected EVD cases, and malaria testing is recommended to guide further actions [12]. Treatment with broad spectrum antibiotics is recommended especially in children, aged less than 5 years [12] as empiric treatment for possible bacterial sepsis. Fluid replacement will be required in many EVD patients and blood transfusions, in those with bleeding, may be beneficial [12].

The Ebola virus has been detected in breast milk up to 9 months after infection, and therefore it is not recommended that woman diagnosed with EVD breastfeed [12, 69].

Historically the use of hyperimmune serum for treatment of EVD, but also other VHF cases, has been reported [70–72]. During the West Africa outbreak, the WHO also approved the use of hyperimmune serum for the treatment of patients, although reportedly not widely used [73]. A limited clinical trial during the outbreak indicated the safety, acceptability and feasibility of treatment with hyperimmune sera [74]. Van Griensven and co-authors [74] also argued the scalability of such a product, particularly given the number of potential immune donors now available in Guinea, Liberia and Sierra Leone.

During the West Africa outbreak, experimental drug treatment of patients was also reported. Palich and co-workers [67] reported the survival of a 6 year old child diagnosed with EVD after treating with favipiravir (or T-705) until virus was cleared from his blood. The use of favipiravir in children older than 1 year has been suggested by Bouazza and co-workers [75], but data from ongoing trails are not yet available.

Other treatment option developments have focused on the use of antiviral drugs (such as favipiravir), type I interferon, anti-coagulant drugs, small interfering RNA, monoclonal antibodies and morpholino-oligomers [76–78]. A number of experimental vaccines have undergone safety and immunogenicity studies *in-vivo* in various populations, although children have largely not been included in these studies. In phase 3 studies using a ring vaccination approach for post exposure prevention of EBV disease in potential contacts (>18 years of age) a recombinant vesicular stomatitis virus based vaccine (rVSV-ZEBOV) showed very high efficacy [79]. Social mobilization, patient isolation, infection control and use of personal protection equipment by health workers, contact tracing and monitoring for infection and safe burials are key traditional interventions in outbreaks.

5.2 Arboviral Diseases

Arboviruses are insect-transmitted viruses belonging to the *Flaviviridae*, *Togaviridae* and *Bunyaviridae* virus families. Hematophagous vectors such as mosquitos, ticks, midges and sandflies transmit the viruses to humans and other susceptible vertebrate hosts. More than 500 arboviruses have been catalogued worldwide, with a little more than a quarter of them reported to cause human disease [80]. The clinical spectrum of arboviral disease is vast ranging from asymptomatic infection to life-threatening febrile and hemorrhagic fever. Arboviral disease typically falls within one of four groups of clinical presentation, namely non-specific systemic febrile disease, encephalitis, polyarthralgia or hemorrhagic fever [81]. Most of the arboviruses are however, not limited to causing only one clinical syndrome. West Nile virus may for example cause unapparent infection in up to 80% of infected individuals but may also cause debilitating encephalitis in others [82]. The distribution of specific arboviruses is geographically defined based on insect vector distribution. Several instances of expanding vector distribution, for example, presumptively due to global climate changes, changes in land use patterns and global transportation have been increasingly linked with the emergence of arboviruses worldwide [83, 84]. Examples include the introduction of West Nile virus to North America in 1999 [85, 86], the re-emergence of dengue in South East Asia, South America and Sub-Sahara Africa [87], the emergence of chikungunya in Indian Ocean Islands and the Caribbean [88], and the emergence of Zika virus (ZIKV) in Latin America in 2014 [89].

Clinical diagnosis of arboviral disease is complicated, and under- or misdiagnosis related to the difficulty of clinical diagnosis in the absence of overt complications is common. In addition, the geographical spread of many arboviruses overlap and have similar clinical manifestations relating to complexity in clinical diagnosis [90]. Specialized diagnostic testing is required to confirm a suspicion of arboviral disease, but is not accessible in many countries where it is probably needed the most. Although RT-PCR detection or virus isolation provide definitive diagnosis, laboratory investigations for arboviral infections mostly rely on serological testing and interpretation due to the typically short viremic period associated with most arboviral infections. Interpretation of serological testing for arboviruses is also challenging and confounded by substantial antigenic cross-reaction between viruses belonging to the same family. The cross-reaction phenomenon particularly applies to flavivirus infections and may be further complicated by possible prior infection with a virus from the same family, or a vaccination for yellow fever or Japanese encephalitis.

For the purpose of this chapter the remainder of the discussion will focus on the newly emerging ZIKV.

5.2.1 Epidemiological Aspects of ZIKV

The WHO declared the ZIKV outbreak in the Americas a Public Health Emergency of International Concern [91]. This followed the rapid emergence of the virus in Brazil and virtually the entire South American subcontinent and the recognition of the potential link in the rise of the number of babies born with microcephaly and other neurological abnormalities observed [92].

The ZIKV was reported for the first time following yellow fever research conducted in the Zika Forest of Uganda in 1947 when it was isolated from a febrile rhesus monkey [93]. It was recognized as a disease causing agent in humans in 1954 when diagnosed in febrile patients in Nigeria [94]. In the years to follow, ZIKV remained mostly a research curiosity with a handful of studies that described its ecology and sero-prevalence in selected human populations and it caused sporadic cases and limited outbreaks without reported mortality or fetal effects [94–99] in a number of countries in Africa and Southeast Asia. In 2007, the first major outbreak of ZIKV was reported from the Yap Islands of Micronesia where more than 70% of the Island's population was affected [100–102]. This was followed by an outbreak in French Polynesia in 2013 and 2014, also affecting nearly 70% of the population [103–106]. The virus continued to spread westward, and was reported in Chile's Easter Island in February 2014 [107]. This was followed by detection of the virus in Brazil in May 2015, although molecular studies have suggested that the virus may have been introduced as early as late 2014 [108–110]. By September 2016, ZIKV was reported from more than 47 countries in Latin America and the Caribbean [111]. In addition, probably related to increased vigilance for the virus, outbreaks of the disease have been reported from Micronesia, Western Samoa, Figi, Singapore, Tonga, Samoa and the Marshal Islands, with possible endemic transmission reported in a dozen other countries in Southeast Asia and the Pacific since 2015 [111]. Although, these reports may relate to endemic occurrence of the virus, the possibility of autochthonous outbreaks due to translocation of virus from Latin America is under investigation for some of these outbreaks [111].

The ZIKV, a flavivirus, is transmitted to humans primarily through the bite of the female, day-feeding *Aedes aegypti* mosquito [112]. The role of other mosquito species in transmission of the virus to humans is disputed [113–121]. *Aedes* mosquitoes usually thrive during warm and wet seasons in the tropics and subtropics, with *Aedes aegypti* particularly well suited for breeding in urban settings, hence the risk of transmission of the virus from infected mosquitos to humans. Although, direct mosquito transmission is mostly implicated in transmission of the ZIKV to humans, transmission in-utero, intrapartum and through sexual transmission has been reported [111, 122–133]. Transmission in-utero has been most alarming with the probable link to congenital malformations and microcephaly (see Sect. 5.2.2). Rare cases of transmission through blood transfusion and laboratory exposures have also been reported [105, 134].

5.2.2 Clinical Aspects of ZIKV

Following the bite of an infected mosquito, an incubation period of up to 2 weeks follows, after which some infected individuals will develop the signs and symptoms of ZIKV infection [100]. An estimated 80% of humans becoming infected with the virus will not develop any clinical signs of disease. In the remainder, disease is typically acute but mild and self-resolving [100, 135]. This is marked by a low-grade fever (typically not above 38.5 °C (101.3 °F)), arthralgia, a maculopapular rash, conjunctivitis, headache and asthenia, much like dengue fever [100, 135]. Clinical diagnosis of ZIKV infection is complicated due to its non-specific presentation, but also due to the co-circulation of other arboviruses such as dengue and chikungunya viruses which cause clinically indistinguishable disease in most cases. Co-infection of ZIKV and dengue or chikungunya virus has also been reported [90, 136–138]. This is not entirely surprising given that these viruses are spread by the same species of mosquitoes. Additionally, other infectious etiologies may also present with similar clinical syndrome including parvovirus infection, rubella, measles, leptospirosis, malaria and rickettsiosis [139].

Hospitalization is rarely required for ZIKV infection [103, 140, 141]. The case fatality rate of ZIKV is considered low and probably associated with comorbidities in such cases [142]. Infants and children present with similar disease as adults and recovery with no sequelae noted in such cases [100, 140, 143, 144].

5.2.2.1 Complications of ZIKV Infection

Neurological complications have been implicated as untoward effects of ZIKV infection. Such neurological deficits have also been reported with other arboviral infections, albeit rarely [145–147]. Guillian-Barré syndrome has been reported as a rare complication of ZIKV infection, although the true extent of the association and occurrence remain to be determined [103, 141, 148]. In addition, rare cases of brain ischemia, encephalitis and myelitis have been reported following ZIKV infection [149–152].

A place and time association was noted with the outbreak of ZIKV in Brazil and concomitant rise in number of babies born with microcephaly in 2015 [153], but studies have subsequently reported significant evidence to support this hypothesis [154]. As such, current opinion is that ZIKV infection during pregnancy can be linked to microcephaly and other congenital neurological and developmental abnormalities, although Koch's postulates have not been applied to prove causality [155]. Reports of neonates with microcephaly and other malformations and abnormalities are accumulating since early 2016 [125, 130, 131, 156–158]. In conjunction with microcephaly, small brain size, ventriculomegaly, hypoplasia or agyria, agenesis of the cerebellar vermis, posterior fossa abnormalities and calcifications have been reported [125, 131, 156, 157]. Other abnormalities have included growth retardation, arthogryposis and ophthalmologic findings [125, 130–132, 156, 157, 159]. Fetal losses have also been reported, and a perinatal mortality rate of almost 27% was reported by Oliveira and co-workers

[130, 131] in a small cohort study conducted in Brazil. Continued longitudinal monitoring of the development of the birth cohort affected by the ZIKV outbreak will reveal the breadth of neurological damage possibly caused by the virus [160]. A retrospective study has indicated cases of microcephaly and other abnormalities in babies born during the ZIKV outbreak in French Polynesia [161]. The possible link between these cases and ZIKV infection was not made at the time, most likely due to the limited number of cases involved (i.e. the extent of the outbreak in Brazil and South America allowed for the possible correlation to be made).

The risk of the development of fetal abnormalities during pregnancy remains to be fully elucidated. The risk factors and the risk associated with infection at different stages of gestation are not clear yet [162]. It is suggested that infection during the first trimester is the most damaging, although infection at later stages of gestation have also reportedly resulted in malformations and neurodevopmental problems [162].

5.2.3   Laboratory Diagnosis of ZIKV

Specialized laboratory testing is required for confirmation of a ZIKV diagnosis. In some settings this service is provided by national reference laboratories, but with the extent of the problem in Latin America the growing availability of commercially available reagents, these tests are offered by an increasing number of laboratories. Testing is recommended for symptomatic individuals with possible exposure to the virus [163]. Asymptomatic pregnant woman at risk of infection should also be screened for possible infection [163].

It is recommended that laboratory tests for ZIKV infection include RT-PCR detection of the viral RNA in whole blood, serum, plasma or urine which are collected during the acute phase of illness (i.e. <7 days after onset of illness) [163, 164]. The virus has also been detected in saliva, cerebrospinal fluid, female genital tract secretions, semen and breast milk [122, 128, 129, 161, 165–169]. Detection of viral RNA in secretions such as semen and saliva has been noted for extended periods even after viremia in the blood has been cleared [122, 129, 168, 170]. Although viremia in the blood is typically cleared quickly, viral RNA has been detected in blood more than 2 months after onset of illness in a limited number of cases [126, 171].

Detection of ZIKV-specific IgM responses requires verification with plaque reduction assays in order to determine the specificity of such responses and mitigate the risk of erroneous diagnosis due to the high level of cross reactivity expected for the antibody responses to other flaviviruses or flavivirus vaccines [163].

Fetal infection has been confirmed by RT-PCR on amniotic fluid, placenta and fetal serum, cerebrospinal fluid and brain [125, 130, 131, 156, 157]. Immunohistochemistry and electron microscopy of brain tissue has also been reported.

5.2.4 Treatment and Prevention of ZIKV

No specific treatment is yet available for ZIKV infection. Treatment is supportive and constitutes bed rest, oral hydration, antipyretics and pain medication [164]. Aspirin should not be used to manage pain and fever in children under the age of 12 due to the risk of Reye's syndrome. Pregnant woman, symptomatic or asymptomatic, that may have had exposure to the virus should be screened for the infection [163]. Fetal monitoring including amniocentesis after 21 weeks of gestation and 6 weeks after exposure for microcephaly and other malformations is recommended [149, 164, 172].

No vaccines are currently available for ZIKV infection but current progress is encouraging [173]. Various approaches to the development of such a vaccine are currently underway, with more than 15 such projects initiated by June 2016 [174]. Prevention of infection in the absence of vaccination relies on preventing exposure to the mosquitos that may transmit the ZIKV. This may be achieved by for example applying insect repellents and staying indoors behind screened windows during peak mosquito activity. Many countries have issued travel warnings to pregnant woman, woman planning to become pregnant and their partners not to travel to ZIKV affected areas. Affected countries have advised woman to delay pregnancy at the time of the ZIKV outbreak, which is a difficult approach in countries where up to half of pregnancies are unplanned [175, 176].

5.3 Rabies

Rabies is the most fatal viral infection in recorded history, and although controllable in certain animal populations through vaccine programmes and largely preventable in humans through prevention of exposures and post exposure prophylaxis, clinical rabies disease cannot be treated [177–179]. Hampson et al. [180] estimated that 59,000 human cases of rabies occur in dog rabies affected countries annually. Dog rabies is most commonly reported from developing countries and typically affects poor communities where dog rabies control measures are inadequate. In addition, many epidemiological reports have indicated that children are particularly affected [181–184]. Human cases are underreported in many developing countries due to the lack of specialized diagnostic services to investigate such cases, but also due to the complexities of clinical diagnosis. Consequently, rabies remains an under-recognized and neglected disease in much of the world [180, 185].

The disease is marked by a progressive acute phase culminating in coma and followed by death within a short period in nearly all patients [186, 187]. The disease is caused by viruses belonging to the genus *Lyssavirus* of the family *Rhabdoviridae*. The lyssaviruses represent a diverse group of viruses with the International Committee on the Taxonomy of Viruses recognizing rabies lyssavirus and 13 rabies-related viral species in 2016 [188]. Additional lyssaviruses have been described but not classified taxonomically [189]. Not all of the lyssaviruses have been associated with human cases, and by far the public health burden associated with rabies relates to infection with "classic" rabies viruses (previously denoted as genotype 1 viruses) (RABV) [180, 190]. The RABV circulates in diverse terrestrial mammals which vary by geographical location [185]. The public health burden associated with rabies however relates to rabies in domestic dogs since these animals serve as the most common vector of the disease to the human population [180, 185, 191, 192]. Therefore, most human rabies cases are reported from developing countries where dog rabies control efforts are suboptimal [180, 185].

5.3.1 Epidemiological Aspects

The occurrence of lyssaviruses has been reported virtually globally, with the exception of the poles and some islands [193]. Only one of the currently described lyssaviruses, namely the Mokola virus, has not been associated with a bat reservoir [193, 194]. Nonetheless, only single human cases of rabies have been associated with bat exposures related to European bat lyssavirus 1 and 2; Australian bat virus; Duvenhage virus and Irkut virus [190, 195]. No human cases have been associated with Aravan, Bokeloh bat; Ikoma; Khujand, Shimoni bat, West Caucasian bat lyssavirus or the unclassified Lleida Bat virus to date [196].

Uniquely, rabies virus is associated with insectivorous bat reservoirs in the Americas—this phenomenon has not been reported elsewhere [197–199]. In addition to bats, cycles of rabies are also reported in various terrestrial mammals in the Americas, for example in raccoons and skunks in the USA. Domestic dog rabies is reported mostly from developing countries in Asia and Africa [180], with elimination achieved in the US and various European countries. The role of wildlife rabies varies from region to region. It should however be noted that all mammalian species are susceptible to rabies and may serve as potential vectors of the disease and this should be considered when evaluating possible exposure.

5.3.2 Clinical Aspects

The virus is transmitted in the saliva of an infected animal. Bites, but any small nicks or scrapes or scratches that break the skin, may constitute exposures. Mucosal exposures, such as licking over the face with contact with the nose, eyes and mouth are also considered exposures. Incubation periods range from 4 days to several years and tend to be shortest with severe bites on the face, head and neck, especially in children [186, 187]. The average incubation period for the RABV is 20–90 days with rare cases reported with an extended incubation period [200, 201]. There are no diagnostic tests possible during this period to predict the likelihood of rabies disease following on a potential exposure. A wide range of non-specific prodromal features including fever, headache, myalgia, fatigue, and change in mood and itching at the initial rabies exposure site may be reported, and in the absence of a report of possible exposure to rabies the disease is difficult to diagnose clinically [186].

Human rabies can take two clinical forms [186, 187]:

- The more common, furious or agitated rabies which is characterized by hydrophobia, aerophobia and periods of extreme excitement alternating with lucid intervals, features of autonomic system dysfunction and finally by unconsciousness and complete paralysis. Initially spasms affect mainly the inspiratory muscles. Autonomic nervous system problems are frequently reported
- Dumb or paralytic rabies which affects the medulla, spinal cord and spinal nerves. Paralytic rabies is less common and more difficult to recognize as rabies. Neurological sign include quadriparesis which predominantly involves the proximal muscles, loss of reflexes and paralysis. Cranial nerve involvement may result in ptosis and external ophthalmoplegia. The course of the disease is less acute and even without intensive care patients may survive for up to 30 days before they succumb to bulbar and respiratory paralysis.

The fatal outcome of RABV infection is associated with a progressive encephalitis and eventual multi-organ failure [187, 202]. The differential diagnosis of rabies is broad. A detailed history of previous potential animal exposure is very important, but can also be inaccurate as animal contact may have been forgotten or unnoticed especially as often reported where bats have been involved in the exposures [203]. The differential diagnosis includes other viral causes of encephalitis and bacterial meningitis and cerebral malaria [182, 186]. Tetanus is important in the differential diagnosis and may also be linked to the same animal source, typically there is a short incubation period and the muscle rigidity is constant, without relaxation between spasms. Paralytic rabies can also be confused with poliomyelitis and Guillian-Barré syndrome [204, 205]. Drug intoxications, psychiatric disorders also form part of the differential diagnosis.

5.3.3 Laboratory Diagnosis

Specialized testing is required to confirm a clinical diagnosis of rabies. Antemortem demonstration of virus in saliva or nuchal skin biopsies may be helpful in some cases, but demonstration of virus using molecular or immunofluorescence techniques on post mortem brain tissue remains the gold standard. Most often these tests are provided only by national reference laboratories, but remain unavailable in many countries where human rabies cases still occur. Routine blood screens are not informative for rabies diagnosis and are expected to be within normal ranges [186]. Examination of cerebrospinal fluid may be unremarkable and is not specific. Magnetic-resonance imaging and computed topography may be useful for differential diagnosis, but no remarkable findings are expected for rabies [187].

Ante-mortem diagnosis of rabies can be confirmed through RT-PCR detection of viral RNA in saliva, nuchal biopsies and cerebrospinal fluid [8, 206, 207]. Detection of viral RNA in tears, urine and respiratory secretions has also been reported [208]. Serological investigations are not always useful to confirm a rabies diagnosis, but the detection of virus neutralizing antibodies in the serum of a previously unvaccinated person or in cerebrospinal fluid are considered confirmatory findings [8]. The gold standard for rabies confirmation in humans (and animals) remains examination of brain tissue post mortem by PCR or immunofluorescence [8, 206, 207].

Every attempt should be made to confirm the diagnosis of human rabies as this provides critical information regarding the burden of disease, failed prevention and guides and stimulates improved rabies control in animals.

5.3.4 Treatment and Prevention

Clinical rabies disease remains untreatable and is limited essentially to palliative care through pain relief and sedation [186, 187]. Provision of intravenous morphine and benzodiazepines has been recommended [8, 187]. If the diagnosis of rabies is confirmed emotional support of the patient and their families is important. Standard infection control precautions are applied in the management of a suspected rabies patient. Transmission of the virus from patient to health care workers or family members has not been previously documented [8, 209]. Rabies post exposure prophylaxis (see below) should be provided to individuals bitten by an infected patient or who experience a mucosal exposure to patient saliva [8, 209].

Survival in an unvaccinated patient with rabies has been reported following the experimental treatment of a teenager in the USA in 2004 [210]. The patient, an adolescent girl, was exposed to rabies virus through contact with a bat, and was treated through the induction of coma. Remarkably, the patient mounted a potent natural immune response to the infection, an observation not commonly noted in unvaccinated rabies patients [210]. The protocol has since been repeated in a number of patients with discouraging results [177, 178, 211, 212]. Effective therapeutic interventions still remain to be expounded [178].

As humans are mostly exposed to the RABV through contact with domestic dogs, the vaccination of dogs presents a crucial intervention for prevention of the disease in humans [213, 214]. The latter should be complimented by promotion of responsible animal ownership and bite prevention programmes. Ensuring the availability and accessibility of post-exposure prophylaxis (PEP) for potentially exposed humans is key in the prevention of human rabies. Rabies PEP is highly effective in preventing rabies in persons exposed to RABV provided timeous administration of biologicals and recommended regimens are followed [8]. True failures of PEP are very rare [215, 216]. Modern rabies vaccines are highly purified, cell-derived inactivated vaccines with agreeable safety profiles, very few adverse reactions and are highly immunogenic [8]. Wound cleaning and PEP, done as soon as possible after suspected contact with a rabid animal can prevent the onset of rabies in virtually 100% of exposures. Intensive wound cleaning is a critical, affordable and manageable component of rabies prevention especially in communities without access to rabies biological prevention and may play some role in some cases in preventing progression to rabies disease. Management of the wound includes cleaning with copious amounts of water (even for small wounds) and soap. The wound should be disinfected with 70% ethanol and iodine. Wound management should also include the use of antibiotics and provision of tetanus vaccination/booster vaccination as clinically required.

Rabies PEP should start as soon as possible following on an exposure to a suspected rabid animal. Even if there is a delay between exposure and visit to the healthcare facility, there is no cut off point after which it is too late to initiate rabies PEP. The decision to give PEP or not should also be made on the risk of RABV transmission, for example taking into account: the animal species, provoked versus unprovoked attack, animal behavior, animal health and category of exposure and history of rabies vaccination of the animal if possible and reliable. Vaccination is either through the intramuscular or intradermal route in accordance with WHO recommended regimens [8]. In wounds that drew any amount of blood (including small wounds such as scrapes that broke the skin) (also called category III exposures), rabies immunoglobulin (RIG) is essential. The latter provides passive immunity whilst immunity is mounted against the rabies vaccination. The use of RIG is crucial in prevention of rabies in category III exposures and failure to provide has culminated in prophylaxis failures [217]. Human or equine-derived rabies immunoglobulins are highly effective and generally safe to use [8]. It should be noted that the use of equine derived immunoglobulin should be used in facilities that are able to manage potential severe allergic reactions [8]. The RIG is infiltrated locally into wound site as much as possible. If injection locally into wound is painful, particularly in children; especially if the wound is on the head, face or neck, consider some sedation provided it can be done safely. The costs of RIG, and limited access and availability in many developing countries, limit the success of PEP in developing countries.

Pre-exposure vaccination is provided to individuals at high or continual risk of RABV exposure, such as veterinarians and laboratory workers [8]. Rabies vaccination is also provided to travellers visiting areas that are endemic for rabies and if their activities may put them at risk of potential exposure (for example only visiting hotel versus backpacking in the outdoors) [218–220]. Pre-exposure regimens using cell-derived vaccines either administered intradermally or intramuscularly as three doses have been shown to induce good immune responses, which are long lasting and successfully boosted even after many years if there is specific exposure [8]. Pre-exposure vaccination obviates the need for post exposure rabies immunoglobulin, a generally scarce resource in many areas. The use of pre-exposure vaccines as part of the expanded program for immunization has been used in some areas with high exposure risks and shown to be immunogenic and safe, with persistence of rabies virus-neutralizing antibodies [221–223].

5.4   Conclusion

Emerging and vector-borne and neglected viral diseases pose a risk to communities particularly in resource poor settings. While adults are disproportionately affected by some of these as a result of occupational exposure, the burden of rabies and Zika infections is borne by children and the fetus respectively. Treatment options overall are very limited or not possible, so early identification of emergence and prevention of spread, are key strategies. The development of vaccines for a number of emerging diseases show promise but clinical trials are still needed. A "One Health" approach addressing environmental, animal and human aspects of emerging and zoonotic diseases are key in prevention, early detection and control.

References

1. Fhogartaigh CN, Aarons E. Viral haemorrhagic fever. Clin Med. 2015;15(1):61–6.
2. Fletcher TE, Brooks TJ, Beeching NJ. Ebola and other viral haemorrhagic fevers. BMJ. 2014;349:g5079.
3. Colebunders R, Van Esbroeck M, Moreau M, Borchert M. Imported viral haemorrhagic fever with a potential for person-to-person transmission: review and recommendations for initial management of a suspected case in Belgium. Acta Clin Belg. 2002;57(5):233–40.
4. Ftika L, Maltezou HC. Viral haemorrhagic fevers in healthcare settings. J Hosp Infect. 2013;83(3):185–92.
5. Iroezindu MO, Unigwe US, Okwara CC, et al. Lessons learnt from the management of a case of Lassa fever and follow-up of nosocomial primary contacts in Nigeria during Ebola virus disease outbreak in West Africa. Trop Med Intern Health. 2015;20(11):1424–30.
6. Moore LS, Moore M, Sriskandan S. Ebola and other viral haemorrhagic fevers: a local operational approach. Br J Med. 2014;75(9):515–22.
7. Roy KM, Ahmed S, Inkster T, et al. Managing the risk of viral haemorrhagic fever transmission in a non-high-level intensive care unit: experiences from a case of Crimean-Congo haemorrhagic fever in Scotland. J Hosp Infect. 2016;93(3):304–8.
8. Vaccines and vaccination against yellow fever. WHO position paper—June 2013. Wkly Epidemiol Rec 2013;88(27):269–83.
9. Asogun AD, Ochei O, Moody V, et al. Clinical presentation and outcome of ribavirin treated RT-PCR confirmed Lassa fever patients in ISTH Irrua: a pilot study. Res Health Sci. 2016;1(2):68–77.
10. Hartnett JN, Boisen ML, Oottamasathien D, et al. Current and emerging strategies for the diagnosis, prevention and treatment of Lassa fever. Future Virol. 2015;10(5):559–84.
11. Ozbey SB, Kader Ç, Erbay A, Ergönül Ö. Early use of ribavirin is beneficial in Crimean-Congo hemorrhagic fever. Vect Zoonotic Dis. 2014;14(4):300–2.
12. World Health Organization. Clinical management of patients with viral haemorrhagic fever: a pocket guide for front-line health workers: interim emergency guidance for country adaptation. Geneva: WHO; 2016.
13. Uyar Y, Christova I, Papa A. Current situation of Crimean Congo hemorrhagic fever (CCHF) in Anatolia and Balkan Peninsula. Tűrk Higj Den Biyol Derg. 2011;68(3):139–51.
14. Ergönül Ö. Crimean–Congo hemorrhagic fever virus: new outbreaks, new discoveries. Curr Opin Virol. 2012;2(2):215–20.
15. Bente DA, Forrester NL, Watts DM, et al. Crimean-Congo hemorrhagic fever: history, epidemiology, pathogenesis, clinical syndrome and genetic diversity. Antiviral Res. 2013;100(1):159–89.
16. Mardani M, Pourkaveh B. Crimean-Congo hemorrhagic fever. Arch Clin Infect Dis. 2013; 7(1):36–42.
17. Elaldi N, Kaya S. Crimean-Congo hemorrhagic fever. J Microbiol Infect Dis. 2014; 4(5):s1–9.
18. Messina JP, Pigott DM, Duda KA, et al. A global compendium of human Crimean-Congo haemorrhagic fever virus occurrence. Sci Data. 2015;2:150016.
19. Shayan S, Bokaean M, Shahrivar MR, Chinikar S. Crimean-Congo Hemorrhagic Fever. Lab Med. 2015;46(3):180–9.

20. Leblebicioglu H, Ozaras R, Fletcher TE, Beeching NJ. Crimean-Congo haemorrhagic fever in travellers: a systematic review. Travel Med Infect Dis. 2016;14(2):73–80.
21. Brauburger K, Hume AJ, Mühlberger E, Olejnik J. Forty-five years of Marburg virus research. Viruses. 2012;4(10):1878–927.
22. Olival KJ, Hayman DT. Filoviruses in bats: current knowledge and future directions. Viruses. 2014;6(4):1759–88.
23. Brainard J, Hooper L, Pond K, Edmunds K, Hunter PR. Risk factors for transmission of Ebola or Marburg virus disease: a systematic review and meta-analysis. Int J Epidemiol. 2015.
24. Rougeron V, Feldmann H, Grard G, et al. Ebola and Marburg haemorrhagic fever. J Clin Virol. 2015;64:111–9.
25. Richmond JK, Baglole DJ. Clinical review. Lassa fever: epidemiology, clinical features and social consequences. BMJ. 2003;327:1271.
26. Ogbu O, Ajuluchukwu E, Uneke CJ. Lassa fever in West African sub-region. An overview. J Vect Borne Dis. 2007;44(1):1–11.
27. Gardner CL, Ryman KD. Yellow fever: a reemerging threat. Clin Lab Med. 2010;30:237–60.
28. Garske T, Van Kerkhove MD, Yactayo S, et al. Yellow Fever in Africa: estimating the burden of disease and impact of mass vaccination from outbreak and serological data. PLoS Med. 2014;11:e1001638.
29. Weyer J, Grobbelaar A, Blumberg L. Ebola virus disease: history, epidemiology and outbreaks. Curr Infect Dis Rep. 2015;17(5):1–8.
30. Centers for Disease Control and Prevention (2016) Outbreak chronology: Ebola virus disease. http://wwwcdcgov/vhf/ebola/outbreaks/history/chronologyhtml. Accessed 16 Sept 2016
31. Centers for Disease Control and Prevention. Question and answers: Zika virus infection (Zika) and pregnancy. http://www.cdc.gov/zika/pregnancy/question¬answers.html. Accessed 5 Feb 2016
32. Scott FD. Ebola haemorrhagic fever: why were children spared? Paediatr Infect Dis J. 1996;15:189–91.
33. Mupere E, Kaducu OF, Yoti Z. Ebola haemorraghic fever among hospitalized children and adolescents in northern Uganda: epidemiologic and clinical observations. Afr Health Sci. 2001;1(2):60–5.
34. World Health Organization Ebola Response Team. Ebola virus disease among children in West Africa. N Engl J Med. 2015;372:1274–7.
35. Baize S, Pannetier D, Oestereich L, et al. Emergency of Zaire Ebola virus disease in Guinea—preliminary report. N Engl J Med. 2014;371(15):1418–28.
36. Saéz M, Weiss S, Nowak K, et al. Investigating the zoonotic origin of the West Africa Ebola epidemic. EMBO Mol Med. 2014;7(1):17–23.
37. Leendertz SAJ, Gogarten JF, Düx A, et al. Assessing the evidence supporting fruit bats as the primary reservoirs for Ebola viruses. Ecohealth. 2016;13(1):18–25.
38. Leroy EM, Kumulungui B, Pourrut X, et al. Fruit bats as reservoirs of Ebola virus. Nature. 2007;438:575–6.
39. Pourrut X, Délicate A, Rollin P, et al. Spatial and temporal patterns of Zaire ebolavirus antibody prevalence in the possible reservoir bat species. J Infect Dis. 2007;196(2):S176.
40. Mann E, Streng S, Bergeron J, Kircher A. A review of the role of food and the food system in the transmission and spread of Ebola virus. PLoS Negl Trop Dis. 2015;9(12):e0004160.
41. Pourrut X, Kumulungui B, Wittmann T, et al. The natural history of Ebola virus in Africa. Microbes Infect. 2005;7:1005–14.
42. Formenty P. Ebola virus disease. In: Ergönül O, Can F, Akova M, Madoff L, editors. Emerging infectious diseases, clinical case studies. London: Academic; 2014. p. 121–34.
43. Goba A, Khan SH, Fonnie M, et al. An outbreak of Ebola virus disease in the Lassa fever zone. J Infect Dis. 2016.
44. Martines RB, Ng DL, Greer PW, et al. Tissue and cellular tropism, pathology and pathogenesis of Ebola and Marburg viruses. J Pathol. 2015;235(2):153–74.

45. O'Shea MK, Clay KA, Craig DG, et al. Diagnosis of febrile illnesses other than Ebola virus disease at an Ebola treatment unit in Sierra Leone. Clin Infect Dis. 2015;61(5):795–8.
46. Ansumana R, Jacobsen KH, Idris MB, et al. Ebola in Freetown area, Sierra Leone - a case study of 581 patients. N Engl J Med. 2015;372(6):587–8.
47. Barry M, Traoré FA, Sako FB, et al. Ebola outbreak in Conakry, Guinea: epidemiological, clinical, and outcome features. Med Mal Infect. 2014;44(11):491–4.
48. Chertow DS, Kleine C, Edwards JK, et al. Ebola virus disease in West Africa - clinical manifestations and management. N Engl J Med. 2014;371(22):2054–7.
49. Ansari AA. Clinical features and pathobiology of Ebolavirus infection. J Autoimmun. 2014;55:1–9.
50. Bwaka MA, Bonnet MJ, Calain P, et al. Ebola hemorrhagic fever in Kikwit, Democratic Republic of the Congo: clinical observations in 103 patients. J Infect Dis. 1999;179(Suppl 1):S1–7.
51. Olupot-Olupot P. Ebola in children: Epidemiology, clinical features, diagnosis and outcomes. Pediatr Infect Dis J. 2015;34(3):314–6.
52. Bebell LM, Riley LE. Ebola virus disease and Marburg disease in pregnancy: a review and management considerations for filovirus infection. Obstet Gynecol. 2015;125(6):1293–8.
53. Baggi F, Taybi A, Kurth A, et al. Management of pregnant women infected with Ebola virus in a treatment centre in Guinea, June 2014. Eurosurveillance. 2014;19(49):2–5.
54. Broadhurst MJ, Brooks TJ, Pollock NR. Diagnosis of Ebola virus disease: past, present, and future. Clin Microbiol Rev. 2016;29(4):773–93.
55. Chua AC, Cunningham J, Moussy F, et al. The case for improved diagnostic tools to control Ebola virus disease in West Africa and how to get there. PLoS Negl Trop Dis. 2015;9(6):e0003734.
56. Martin P, Laupland KB, Frost EH, Valiquette L. Laboratory diagnosis of Ebola virus disease. Intensive Care Med. 2015;41(5):895–8.
57. Nouvellet P, Garske T, Mills HL, et al. The role of rapid diagnostics in managing Ebola epidemics. Nature. 2015;528(7580):S109–16.
58. Ahrberg CD, Manz A, Neužil P. Palm-sized device for point-of-care Ebola detection. Anal Chem. 2016;88(9):4803–7.
59. Broadhurst MJ, Kelly JD, Miller A, et al. ReEBOV Antigen Rapid Test kit for point-of-care and laboratory-based testing for Ebola virus disease: a field validation study. Lancet. 2015;386(9996):867–74.
60. Kurosaki Y, Magassouba NF, Oloniniyi OK, et al. Development and evaluation of reverse transcription-loop-mediated isothermal amplification (RT-LAMP) assay coupled with a portable device for rapid diagnosis of Ebola virus disease in Guinea. PLoS Negl Trop Dis. 2016;10(2):e0004472.
61. Semper AE, Broadhurst MJ, Richards J, et al. Performance of the GeneXpert Ebola Assay for diagnosis of Ebola virus disease in Sierra Leone: a field evaluation study. PLoS Med. 2016;13(3):e1001980.
62. Walker NF, Brown CS, Youkee D, et al. Evaluation of a point-of-care blood test for identification of Ebola virus disease at Ebola holding units, Western Area, Sierra Leone, January to February 2015. Euro Surveill. 2015;20(12):64.
63. Weller SA, Bailey D, Matthews S, et al. Evaluation of the Biofire FilmArray BioThreat-E Test (v2. 5) for rapid identification of Ebola virus disease in heat-treated blood samples obtained in Sierra Leone and the United Kingdom. J Clin Microbiol. 2016;54(1):114–9.
64. Geisbert TW. Emergency treatment for exposure to Ebola virus: the need to fast-track promising vaccines. JAMA. 2015;313(12):1221–2.
65. Kilgore PE, Grabenstein JD, Salim AM, Rybak M. Treatment of Ebola virus disease. Pharmacotherapy. 2015;35(1):43–53.
66. Li H, Ying T, Yu F, et al. Development of therapeutics for treatment of Ebola virus infection. Microbes Infect. 2015;17(2):109–17.
67. Palich R, Gala JL, Petitjean F, et al. A 6-year-old child with severe Ebola virus disease: laboratory-guided clinical care in an Ebola Treatment Center in Guinea. PLoS Negl Trop Dis. 2016;10(3):e0004393.

68. http://apps.who.int/medicinedocs/documents/s22501en/s22501en.pdf. Accessed 6 Oct 2016
69. Bausch DG. Assessment of the risk of Ebola virus transmission from bodily fluids and fomites. J Infect Dis. 2007;196:S142–7.
70. Kudoyarova-Zubavichene NM, Sergeyev NN, Chepurnov AA, Netesov SV. Preparation and use of hyperimmune serum for prophylaxis and therapy of Ebola virus infections. J Infect Dis. 1999;179(1):S218–23.
71. Maruyama T, Rodriguez LL, Jahrling PB, et al. Ebola virus can be effectively neutralized by antibody produced in natural human infection. J Virol. 1999;73(7):6024–30.
72. Mupapa K, Massamba M, Kibadi K, et al. Treatment of Ebola hemorrhagic fever with blood transfusions from convalescent patients. J Infect Dis. 1999;179(1):S18–23.
73. Gulland A. First Ebola treatment is approved by WHO. BMJ. 2014;349:g5539.
74. Van Griensven J, De Weiggheleire A, Delamou A, et al. The use of Ebola convalescent plasma to treat Ebola virus disease in resource constrained settings: a perspective from the field. Clin Infect Dis. 2015.
75. Bouazza N, Treluyer JM, Foissac F, et al. Favipiravir for children with Ebola. Lancet. 2015;385(9968):603–4.
76. Bishop BM. Potential and emerging treatment options for Ebola virus disease. Ann Pharmacother. 2014;49(2):196–206.
77. Madelain V, Nguyen THT, Olivo A, et al. Ebola virus infection: review of the pharmacokinetic and pharmacodynamic properties of drugs considered for testing in human efficacy trials. Clin Pharmacokinet. 2016;55(8):907–23.
78. Sivanandy P, Sin SH, Ching OY, et al. A review on current trends in the management of Ebola virus disease. Int J Pharm Teaching Pract. 2016;7(1):2657–65.
79. Henao-Restrepo AM, Longini IM, Egger M, et al. Efficacy and effectiveness of an rVSV-vectored vaccine expressing Ebola surface glycoprotein: interim results from the Guinea ring vaccination cluster-randomised trial. Lancet. 2015;386(9996):857–66.
80. Gubler DJ. Human arbovirus infections worldwide. Ann N Y Acad Sci. 2001;951(1):13–24.
81. Alatoom A, Payne D. An overview of arboviruses and bunyaviruses. Lab Med. 2009;40(4):237–40.
82. Hayes EB, Sejvar JJ, Zaki SR, et al. Virology, pathology, and clinical manifestations of West Nile virus disease. Emerg Infect Dis. 2005;11(8):1174–9.
83. Gubler DJ. The global emergence/resurgence of arboviral diseases as public health problems. Arch Med Res. 2002;33(4):330–42.
84. Weaver SC, Barrett AD. Transmission cycles, host range, evolution and emergence of arboviral disease. Nat Rev Microbiol. 2004;2(10):789–801.
85. Garmendia AE, Van Kruiningen HJ, French RA. The West Nile virus: its recent emergence in North America. Microbes Infect. 2001;3(3):223–9.
86. Nash D, Mostashari F, Fine A, et al. The outbreak of West Nile virus infection in the New York City area in 1999. N Engl J Med. 2001;344(24):1807–14.
87. Messina JP, Brady OJ, Scott TW, et al. Global spread of dengue virus types: mapping the 70 year history. Trends Microbiol. 2014;22(3):138–46.
88. Weaver SC. Arrival of chikungunya virus in the new world: prospects for spread and impact on public health. PLoS Negl Trop Dis. 2014;8(6):e2921.
89. Fauci AS, Morens DM. Zika virus in the Americas—yet another arbovirus threat. N Engl J Med. 2016;374(7):601–4.
90. Moulin E, Selby K, Cherpillod P, Kaiser L, Boillat-Bianco N. Simultaneous outbreaks of dengue, chikungunya and Zika virus infections: diagnosis challenge in a returning traveller with nonspecific febrile illness. New Microbes New Infect. 2016;11:6–7.
91. Gulland A. Zika virus is a global public health emergency, declares WHO. BMJ. 2016; 352:i657.
92. Heymann DL, Hodgson A, Freedman DO, et al. Zika virus and microcephaly: why is this situation a PHEIC? Lancet. 2016;387(10020):719–21.
93. Dick GWA, Kitchen SF, Haddow AJ. Zika virus (I). Isolations and serological specificity. Trans R Soc Trop Med Hyg. 1952;46(5):509–20.

94. Fagbami AH. Zika virus infections in Nigeria: virological and seroepidemiological investigations in Oyo State. J Hygiene. 1979;83(02):213–9.
95. Darwish MA, Hoogstraal H, Roberts TJ, et al. A sero-epidemiological survey for certain arboviruses (Togaviridae) in Pakistan. Trans R Soc Trop Med Hyg. 1983;77(4):442–5.
96. Haddow AJ, Williams MC, Woodall JP, et al. Twelve isolations of Zika virus from Aedes (Stegomyia) africanus (Theobald) taken in and above a Uganda forest. Bull World Health Organ. 1964;31(1):57.
97. Macnamara FN. Zika virus: a report on three cases of human infection during an epidemic of jaundice in Nigeria. Trans R Soc Trop Med Hyg. 1954;48(2):139–45.
98. Olson JG, Ksiazek TG. Zika virus, a cause of fever in Central Java, Indonesia. Trans R Soc Trop Med Hyg. 1981;75(3):389–93.
99. Simpson DIH. Zika virus infection in man. Trans R Soc Trop Med Hyg. 1964;58(4):335–8.
100. Duffy MR, Chen TH, Hancock WT, et al. Zika virus outbreak on Yap Island, Federated States of Micronesia. N Engl J Med. 2009;360:2536.
101. Hayes EB. Zika virus outside Africa. Emerg Infect Dis. 2009;15:1347.
102. Lanciotti RS, Kosoy OL, Laven JJ, et al. Genetic and serologic properties of Zika virus associated with an epidemic, Yap State, Micronesia, 2007. Emerg Infect Dis. 2008;14:1232.
103. Cao-Lormeau V-M, Blake A, Mons S, et al. Guillain-Barré Syndrome outbreak associated with Zika virus infection in French Polynesia: a case-control study. Lancet. 2016;387(10027):1531–9.
104. Musso D, Nilles EJ, Cao-Lormeau V-M. Rapid spread of emerging Zika virus in the Pacific area. Clin Microbiol Infect. 2014;20(10):O595–6.
105. Musso D, Nhan T, Robin E, et al. Potential for Zika virus transmission through blood transfusion demonstrated during an outbreak in French Polynesia, November 2013 to February 2014. Euro Surveill. 2014;19(14):20761.
106. Cao-Lormeau V-M, Roche C, et al. Zika virus: French Polynesia, South Pacific, 2013. EID. 2014;20(6):1085–6.
107. Tognarelli J, Ulloa S, Villagra E, Lagos J, Aguayo C, Fasce R, Parra B, Mora J, Becerra N, Lagos N, Vera L. A report on the outbreak of Zika virus on Easter Island, South Pacific, 2014. Arch Virol. 2016;161(3):665–8.
108. Campos GS, Bandeira AC, Sardi SI. Zika virus outbreak, Bahia, Brazil. Emerg Infect Dis. 2015;21(10):1885.
109. Faria NR, Azevedo Rdo S, Kraemer MU, et al. Zika virus in the Americas: early epidemiological and genetic findings. Science. 2016;352:345.
110. Zanluca C, Melo VCAD, Mosimann ALP, Santos GIVD, Santos CNDD, Luz K. First report of autochthonous transmission of Zika virus in Brazil. Mem Inst Oswaldo Cruz. 2015;110(4): 569–72.
111. World Health Organization (2016). Zika virus. Microcephaly Guillian-Barré syndrome situation report 29 September 2016. http://apps.who.int/iris/bitstream/10665/250244/1/zikasitrep29Sep16-eng.pdf?ua=1. Accessed 6 Oct 2016.
112. Gubler D, Kuno G, Markoff L. Flaviviruses. In: Knipe DM, Howley PM, editors. Fields virology. Philadelphia: Lippincott, Williams and Wilkins; 2007. p. 1153–252.
113. Chouin-Carneiro T, Vega-Rua A, Vazeille M, et al. Differential susceptibilities of Aedes aegypti and Aedes albopictus from the Americas to Zika Virus. PLoS Negl Trop Dis. 2016;10(3):e0004543.
114. Diagne CT, Diallo D, Faye O, et al. Potential of selected Senegalese Aedes spp. mosquitoes (Diptera: Culicidae) to transmit Zika virus. BMC Infect Dis. 2015;15(1):1.
115. Diallo D, Sall AA, Diagne CT, et al. Zika virus emergence in mosquitoes in southeastern Senegal, 2011. PLoS One. 2014;9(10):e109442.
116. Faye O, Faye O, Diallo D, et al. Quantitative real-time PCR detection of Zika virus and evaluation with field-caught mosquitoes. Virol J. 2013;10(1):1.
117. Grard G, Caron M, Mombo IM, et al. Zika virus in Gabon (Central Africa) - 2007: a new threat from Aedes albopictus? PLoS Negl Trop Dis. 2014;8(2):e2681.
118. Ledermann JP, Guillaumot L, Yug L, et al. Aedes hensilli as a potential vector of Chikungunya and Zika viruses. PLoS Negl Trop Dis. 2014;8(10):e3188.

119. Li MI, Wong PSJ, Ng LC, Tan CH. Oral susceptibility of Singapore *Aedes* (*Stegomyia*) *aegypti* (*Linnaeus*) to Zika virus. PLoS Negl Trop Dis. 2012;6(8):e1792.
120. Marcondes CB, Ximenes MDFFD. Zika virus in Brazil and the danger of infestation by *Aedes (Stegomyia)* mosquitoes. Rev Soc Bras Med Trop. 2016;49(1):4–10.
121. Wong PSJ, Li MZI, Chong CS, et al. *Aedes* (*Stegomyia*) *albopictus* (Skuse): a potential vector of Zika virus in Singapore. PLoS Negl Trop Dis. 2013;7(8):e2348.
122. Atkinson B, Hearn P, Afrough B, et al. Detection of Zika virus in semen. Emerg Infect Dis. 2016;22(5):940.
123. Besnard M, Lastere S, Teissier A, Cao-Lormeau VM, Musso D. Evidence of perinatal transmission of Zika virus, French Polynesia, December 2013 and February 2014. Euro Surveill. 2014;19(13):20751.
124. Brooks JT. Update: interim guidance for prevention of sexual transmission of Zika virus - United States, July 2016. MMWR Morb Mortal Wkly Rep. 2016;65:120–1.
125. Calvet G, Aguiar RS, Melo AS, Sampaio SA, de Filippis I, Fabri A, Araujo ES, de Sequeira PC, de Mendonça MC, de Oliveira L, Tschoeke DA. Detection and sequencing of Zika virus from amniotic fluid of fetuses with microcephaly in Brazil: a case study. Lancet Infect Dis. 2016;16:653–60.
126. Driggers RW, Ho CY, Korhonen EM, et al. Zika virus infection with prolonged maternal viremia and fetal brain abnormalities. N Engl J Med. 2016;374(22):2142–51.
127. Foy BD, Kobylinski KC, Chilson Foy JL, Blitvich BJ, Travassos da Rosa A, Haddow AD, Lanciotti RS, Tesh RB. Probable non-vector-borne transmission of Zika virus, Colorado, USA. Emerg Infect Dis. 2011;17(5):880–2.
128. Musso D, Roche C, Nhan TX, et al. Detection of Zika virus in saliva. J Clin Virol. 2015;68:53–5.
129. Musso D, Roche C, Robin E, et al. Potential sexual transmission of Zika virus. Emerg Infect Dis. 2015;21(2):359–61.
130. Oliveira MAS, Aguiar RS, Ramos Amorim MR, et al. Congenital zika virus infection: beyond neonatal microcephaly. JAMA Neurol. 2016.
131. Oliveira MAS, Malinger G, Ximenes R, et al. Zika virus intrauterine infection causes fetal brain abnormality and microcephaly: tip of the iceberg? Ultrasound Obstet Gynecol. 2016; 47(1):6–7.
132. Sarno M, Sacramento GA, Khouri R, do Rosário MS, Costa F, Archanjo G, Santos LA, Nery Jr N, Vasilakis N, Ko AI, de Almeida AR. Zika virus infection and stillbirths: a case of hydrops fetalis, hydranencephaly and fetal demise. PLoS Negl Trop Dis. 2016;10(2):e0004517.
133. Venturi G, Zammarchi L, Fortuna C, Remoli ME, Benedetti E, Fiorentini C, Trotta M, Rizzo C, Mantella A, Rezza G, Bartoloni A. An autochthonous case of Zika due to possible sexual transmission, Florence, Italy, 2014. Euro Surveill. 2016;21(8):30148.
134. Filipe AR, Martins CMV, Rocha H. Laboratory infection with Zika virus after vaccination against yellow fever. Arch Gesamte Virusforsch. 1973;43(4):315–9.
135. Brasil P, Calvet GA, Siqueira AM, Wakimoto M, de Sequeira PC, Nobre A, Quintana MDSB, de Mendonça MCL, Lupi O, de Souza RV, Romero C. Zika virus outbreak in Rio de Janeiro, Brazil: clinical characterization, epidemiological and virological aspects. PLoS Negl Trop Dis. 2016;10(4):e0004636.
136. Dupont-Rouzeyrol M, O'Connor O, Calvez E, et al. Co-infection with Zika and dengue viruses in 2 patients, New Caledonia, 2014. Emerg Infect Dis. 2015;21(2):381–2.
137. Roth A, Mercier A, Lepers C, et al. Concurrent outbreaks of dengue, chikungunya and Zika virus infections-an unprecedented epidemic wave of mosquito-borne viruses in the Pacific 2012-2014. Euro Surveill. 2014;19(41):20929.
138. Villamil-Gómez WE, González-Camargo O, Rodriguez-Ayubi J, et al. Dengue, chikungunya and Zika co-infection in a patient from Colombia. J Infect Public Health. 2016;9(5):684–6.
139. LaBeaud AD (2016) Zika virus infection: an overview. In: Hirsch MS, Baron EL, editors. UptoDate.com Wolters Kluwer. Topic 106169, Version 93.0.
140. Goodman AB, Dziuban EJ, Powell K, et al. Characteristic of children age < 18 years with zika virus disease acquired postnatally—US States, January 2015-July 2016. MMWR Morb Mortal Wkly Rep. 2016;65:1082–5.

141. Oehler E, Watrin L, Larre P, et al. Zika virus infection complicated by Guillain-Barre syndrome—case report, French Polynesia, December 2013. Euro Surveill. 2014;19(9):20720.
142. Sarmiento-Ospina A, Vásquez-Serna H, Jimenez-Canizales CE, et al. Zika virus associated deaths in Colombia. Lancet Infect Dis. 2016;16(5):523–4.
143. Fleming-Dutra KE, Nelson JM, Fischer M, et al. Update: Interim guidelines for Health Care Providers caring for infants and children with possible Zika Virus infection United States. MMWR Morb Mortal Wkly Rep. 2016;65(7):182–7.
144. Karwowski MP, Nelson JM, Staples JE, et al. Zika virus disease: A CDC update for pediatric health care providers. Paediatrics. 2016;137(5):e20160621.
145. Gérardin P, Sampéris S, Ramful D, et al. Neurocognitive outcome of children exposed to perinatal mother-to-child chikungunya virus infection: the CHIMERE cohort study on Reunion Island. PLoS Negl Trop Dis. 2014;8:e2996.
146. O'Leary DR, Kuhn S, Kniss KL, et al. Birth outcomes following West Nile virus infection of pregnant women in the United States: 2003-2004. Paediatrics. 2006;117:e537–45.
147. Sips GJ, Wilschut J, Smit JM. Neuroinvasive flavivirus infections. Rev Med Virol. 2011;22:69–87.
148. Smith DW, Mackenzie J. Zika virus and Guillain-Barré syndrome: another viral cause to add to the list. Lancet. 2016;387:1486–8.
149. Baud D, Van Mieghem T, Musso D, et al. Clinical management of pregnant women exposed to Zika virus. Lancet Infect Dis. 2016;16(5):523.
150. Carteaux G, Maquart M, Bedet A, et al. Zika virus associated with meningoencephalitis. N Engl J Med. 2016;374:1595–6.
151. Mécharles S, Herrmann C, Poullian P, et al. Acute myelitis due to Zika virus infection. Lancet. 2016;387:1481.
152. Rozé B, Najioullah F, Signate A et al (2016) Zika virus detection in cerebrospinal fluid from two patients with encephalopathy, Martinique, February 2016. Euro Surveill 21(16), pii: 30205.
153. Schuler-Faccini L, Ribeiro EM, IML F, et al. Possible association between Zika virus infection and microcephaly—Brazil, 2015. MMWR Morb Mortal Wkly Rep. 2016;65(3):59–62.
154. Rodrigues LC. Microcephaly and Zika virus infection. Lancet. 2016;387(10033):2070–2.
155. Tetro JA. Zika and microcephaly: causation, correlation, or coincidence. Microbes Infect. 2016;18(3):167–8.
156. de Araújo TVB, Rodrigues LC, de Alencar Ximenes RA et al (2016). Association between Zika virus infection and microcephaly in Brazil, January to May, 2016: preliminary report of a case-control study. Lancet Infect Dis. http://dx.doi.org/10.1016/
157. Mlakar J, Korva M, Tul N, et al. Zika virus associated with microcephaly. N Engl J Med. 2016;374(10):951–8.
158. Rubin EJ, Greene MF, Baden LR. Zika virus and microcephaly. N Engl J Med. 2016;374(10):984–5.
159. Jampol LM, Goldstein DA. Zika virus infection and the eye. JAMA Ophthalmol. 2016;134(5):535–6.
160. Barton MA, Salvadori MI. Zika virus and microcephaly. Can Med Assoc J. 2016;188(7):E118–9.
161. Besnard M, Eyrolle-Guignot D, Guillemette-Artur P, et al. Congenital cerebral malformations and dysfunction in fetuses and newborns following the 2013 to 2014 Zika virus epidemic in French Polynesia. Euro Surveill. 2016;21(13):30181.
162. Ticconi C, Pietropolli A, Rezza G. Zika virus infection and pregnancy: what we do and do not know. Pathog Glob Health. 2016.
163. World Health Organization (2016) Laboratory testing for Zika virus infection. Interim guidance. WHO/ZIKV/LAB/16.1. Available from: http://apps.who.int/iris/bitstream/10665/204671/1/WHO_ZIKV_LAB_16.1_eng.pdf. Accessed on 20 Oct 2016.
164. Falcao MB, Cimerman S, Luz KG, et al. Management of infection by the Zika virus. Ann Clin Microbiol Antimicrob. 2016;15:57.
165. Barzon L, Pacenti M, Berto A, et al. Isolation of infectious Zika virus from saliva and prolonged viral RNA shedding in a traveller returning from the Dominican Republic to Italy, January 2016. Euro Surveill. 2016;21(10):Article 1.

166. Dupont-Rouzeyrol M, Biron A, O'Connor O, et al. Infectious Zika viral particles in breast-milk. Lancet. 2016;387(10023):1051.
167. Fréour T, Mirallié S, Hubert B, et al. Sexual transmission of Zika virus in an entirely asymptomatic couple returning from a Zika epidemic area, France, April 2016. Euro Surveill. 2016;21(23):30254.
168. Mansuy JM, Dutertre M, Mengelle C, et al. Zika virus: high infectious viral load in semen, a new sexually transmitted pathogen. Lancet Infect Dis. 2016;16(405):00138–9.
169. Prisant N, Bujan L, Benichou H, et al. Zika virus in the female genital tract. Lancet Infect Dis. 2016;16(9):1000–1.
170. Harrower J, Kiedrzynski T, Baker S, et al. Sexual transmission of Zika virus and persistence in semen, New Zealand, 2016. Emerg Infect Dis. 2016;22(10):1855–7.
171. Lustig Y, Mendelson E, Paran N, et al. Detecton of Zika virus RNA in whole blood in imported Zika virus disease cases up to 2 months after symptom onset, Israel, December 2015-April 2016. Euro Surveill. 2016;21(26):30269.
172. Vouga M, Musso D, Van Mieghem T, Baud D. CDC guidelines for pregnant woman during the Zika virus outbreak. Lancet. 2016;387:843–4.
173. Kennedy RB. Pushing forward with Zika vaccines. EBioMedicine. 2016.
174. Tripp RA, Ross TM. Development of a Zika vaccine. Expert Rev Vaccines. 2016;15(9):1083–5.
175. Ndeffo-Mbah ML, Parpia AS, Galvani AP. Mitigating prenatal zika virus infection in the America. Ann Intern Med. 2016;165:551–9.
176. Sedgh G, Singh S, Hussain R. Intended and unintended pregnancies worldwide in 2012 and recent trends. Stud Fam Plann. 2014;45:301–14.
177. Jackson AC. Therapy of human rabies. In: Jackson AC, editor. Advances in virus research: advances in rabies research. Boston: Academic; 2011. p. 365–75.
178. Jackson AC. Current and future approaches to the therapy of human rabies. Antiviral Res. 2013;99(1):61–7.
179. Warrell MJ, Warrell DA. Rabies: the clinical features, management and prevention of the classic zoonosis. Clin Med. 2015;15(1):78–81.
180. Hampson K, Coudeville L, Lembo T, et al. Estimating the global burden of endemic canine rabies. PLoS Negl Trop Diseases. 2015;9(4):e0003709.
181. Ichhpujani RL, Mala C, Veena M, et al. Epidemiology of animal bites and rabies cases in India. A multicentric study. J Commun Dis. 2008;40(1):27–36.
182. Mallewa M, Fooks AR, Banda D, et al. Rabies encephalitis in malaria-endemic area, Malawi, Africa. Emerg Infect Dis. 2007;13(1):136–9.
183. Sriaroon C, Sriaroon P, Daviratanasilpa S, et al. Retrospective: animal attacks and rabies exposures in Thai children. Travel Med Infect Dis. 2006;4(5):270–4.
184. Schalamon J, Ainoedhofer H, Singer G, et al. Analysis of dog bites in children who are younger than 17 years. Pediatrics. 2006;117(3):e374–9.
185. Taylor LH, Nel LH. Global epidemiology of canine rabies: past, present, and future prospects. Vet Med Res Rep. 2015;6:361–71.
186. Hemachudha T, Laothamatas J, Rupprecht CE. Human rabies: a disease of complex neuropathogenetic mechanisms and diagnostic challenges. Lancet Neurol. 2002;1(2):101–9.
187. Hemachudha T, Ugolini G, Wacharapluesadee S, et al. Human rabies: neuropathogenesis, diagnosis, and management. Lancet Neurol. 2013;12(5):498–513.
188. Afonso CL, Amarasinghe GK, Bányai K, et al. Taxonomy of the order Mononegavirales: update 2016. Arch Virol. 2016;161(8):1–10.
189. Ceballos NA, Morón SV, Berciano JM, et al. Novel lyssavirus in bat, Spain. Emerg Infect Dis. 2013;19(5):793–5.
190. Banyard AC, Evans JS, Luo T, Fooks AR. Lyssaviruses and bats: emergence and zoonotic threat. Viruses. 2014;6(8):2974–90.
191. Overall KL, Love M. Dog bites to humans-demography, epidemiology, injury and risk. J Am Vet Med Assoc. 2001;218(12):1923–34.

192. World Health Organisation. WHO Expert Consultation on Rabies. Second report. World Health Organization technical report series 982. Geneva: WHO; 2013.
193. Kuzmin IV, Rupprecht CE. Bat lyssaviruses. In: Wang L-F, Cowled C, editors. Bats and viruses: a new frontier of emerging infectious diseases. India: Wiley Blackwell; 2015. p. 47–97.
194. Kgaladi J, Wright N, Coertse J, et al. Diversity and epidemiology of Mokola virus. PLoS Negl Trop Dis. 2013;7(10):e2511.
195. Johnson N, Vos A, Freuling C, et al. Human rabies due to lyssavirus infection of bat origin. Vet Microbiol. 2010;142(3-4):151–9.
196. Kuzmin IV. Basic facts about Lyssaviruses. In: Rupprecht CE, Nagarajan T, editors. Current laboratory techiques in rabies diagnosis, research and prevention. New York: Elsevier; 2014. p. 3–24.
197. Davis AD, Gordy PA, Bowen RA. Unique characteristics of bat rabies viruses in big brown bats (*Eptesicus fuscus*). Arch Virol. 2013;158(4):809–20.
198. Johnson N, Aréchiga-Ceballos N, Aguilar-Setien A. Vampire bat rabies: ecology, epidemiology and control. Viruses. 2014;6(5):1911–28.
199. Patyk K, Turmelle A, Blanton JD, Rupprecht CE. Trends in national surveillance data for bat rabies in the United States: 2001–2009. Vector Borne Zoonotic Dis. 2012;12(8):666–73.
200. Grattan-Smith PJ, O'Regan WJ, Ellis PS, et al. Rabies. A second Australian case, with a long incubation period. Med J Aust. 1992;156(9):651–4.
201. Johnson N, Fooks A, McColl K. Reexamination of human rabies case with long incubation, Australia. Emerg Infect Dis. 2008;14(12):1950–2.
202. Fu ZF, Jackson AAC. Neuronal dysfunction and death in rabies virus infection. J Neurovirol. 2005;11(1):101–6.
203. Messenger SL, Smith JS, Rupprecht CE. Emerging epidemiology of bat-associated cryptic cases of rabies in humans in the United States. Clin Infect Dis. 2002;35(6):738–47.
204. Cohen R, Babushkin F, Shapiro M, et al. Paralytic rabies misdiagnosed as Guillain-Barre syndrome in a guest worker: a case report. J Neuroinfect Dis. 2016;7:208.
205. Sheikh KA, Ramos-Alvarez M, Jackson AC, et al. Overlap of pathology in paralytic rabies and axonal Guillain–Barré syndrome. Ann Neurol. 2005;57(5):768–72.
206. Fooks AR, Johnson N, Freuling CM, et al. Emerging technologies for the detection of rabies virus: challenges and hopes in the 21st century. PLoS Negl Trop Dis. 2009;3(9):e530.
207. Hanlon CA, Nadin-Davis SA. Laboratory diagnosis of rabies. In: Jackson A, editor. Rabies: scientific basis of the disease and its management. Third ed. Boston: Academic; 2013. p. 409–59.
208. Wacharapluesadee S, Hemachudha T. Urine samples for rabies RNA detection in the diagnosis of rabies in humans. Clin Infect Dis. 2002;34(6):874–5.
209. Stantic-Pavlinic M. Rabies treatment of health care staff. Swiss Med Wkly. 2002;132: 129–31.
210. Willoughby RE, Tieves KS, Hoffman GM, et al. Survival after treatment of rabies with induction of coma. N Engl J Med. 2005;352(24):2508–14.
211. Hemachudha T, Sunsaneewitayakul B, Desudchit T, et al. Failure of therapeutic coma and ketamine for therapy of human rabies. J Neurovirol. 2006;12(5):407–9.
212. Wilde H, Hemachudha T, Jackson AC. Viewpoint: management of human rabies. Trans R Soc Trop Med Hyg. 2008;102(10):979–82.
213. Cleaveland S, Kaare M, Knobel D, Laurenson MK. Canine vaccination—providing broader benefits for disease control. Vet Microbiol. 2006;117(1):43–50.
214. Cleaveland S, Kaare M, Tiringa P, et al. A dog rabies vaccination campaign in rural Africa: impact on the incidence of dog rabies and human dog-bite injuries. Vaccine. 2003;21(17):1965–73.
215. Rupprecht CE, Gibbons RV. Prophylaxis against rabies. N Engl J Med. 2004;351(25): 2626–35.
216. Wilde H. Failures of post-exposure rabies prophylaxis. Vaccine. 2007;25(44):7605–9.
217. Wilde H, Khawplod P, Hemachudha T, Sitprija V. Postexposure treatment of rabies infection: can it be done without immunoglobulin? Clin Infect Dis. 2002;34:477–80.

218. Christiansen AH, Rodriguez AB, Nielsen J, Cowan SA. Should travellers to rabies-endemic countries be pre-exposure vaccinated? An assessment of post-exposure prophylaxis and pre-exposure prophylaxis given to Danes travelling to rabies-endemic countries 2000–12. Journal of Travel Medicine. 2016;23(4):taw022.
219. LeGuerrier P, Pilon PA, Deshaies D, Allard R. Pre-exposure rabies prophylaxis for the international traveller: a decision analysis. Vaccine. 1996;14(2):167–76.
220. Pandey P, Shlim DR, Cave W, Springer MF. Risk of possible exposure to rabies among tourists and foreign residents in Nepal. J Travel Med. 2002;9(3):127–31.
221. Dodet B, Asian Rabies Expert Bureau (AREB). Report of the sixth AREB meeting, Manila, The Philippines, 10-12 November 2009. Vaccine. 2010;28(19):3265–8.
222. Kamoltham T, Thinyounyong W, Phongchamnaphai P, et al. Pre-exposure rabies vaccination using purified chick embryo cell rabies vaccine intradermally is immunogenic and safe. J Pediatr. 2007;151(2):173–7.
223. Malerczyk C, Vakil HB, Bender W. Rabies pre-exposure vaccination of children with purified chick embryo cell vaccine (PCECV). Hum Vaccin Immunother. 2013;9(7):1454–9.

Diagnosis of Viral Infections

6

Marthi Pretorius and Marietjie Venter

Abstract

Accurate diagnosis of viral infections enhances the ability of the clinician to make decisions on appropriate treatment of patients, evaluate disease progression and prevent misuse of antibiotics. Knowledge of the pathogen involved also allow implementation of infection control and monitoring of success of antiviral treatments that may affect the prognosis of patients. Epidemiological data collected through accurate diagnostics play an important role in public health through identification and control of outbreaks, implementation of appropriate diagnostic tests, vaccination programs and treatment but also to recognize common and emerging pathogens in a community. It is key that the clinician have an understanding of appropriate specimens to send to the laboratory and the value of specific nucleic acid and serological testing for different viral pathogens. Molecular techniques have revolutionized viral diagnoses over the past decade and enhanced both the sensitivity and specificity of tests and the speed by which a diagnosis can be made and new tests be developed. The continued use of serology for viruses with a short viremia, or for chronic infections should however complement these tests. This chapter aims to provide an overview of the available tests, the principles of testing and appropriate tests to select for different viruses and syndromes. Also provided is a glimpse of new developments in diagnostics that may further enhance the capacity to make a conclusive diagnosis in the near future.

M. Pretorius, Ph.D. • M. Venter, Ph.D. (✉)
Centre for Viral Zoonoses, Department of Medical Virology, University of Pretoria, Pretoria, South Africa
e-mail: marietjie.venter.up@gmail.com; marietjie.venter@up.ac.za

© Springer International Publishing AG 2017
R.J. Green (ed.), *Viral Infections in Children, Volume I*,
DOI 10.1007/978-3-319-54033-7_6

6.1   Introduction

Human virus infections may affect all ages and may impact morbidity and mortality through acute, chronic, recurrent or lifelong infections. This may depend on the immune status of the patient and their ability to clear virus infection as well as the characteristics of the pathogen. The development of sensitive and specific methods for both the detection of viral nucleic acids and antiviral antibodies has greatly advanced our ability to make accurate diagnoses at different stages of the disease. Previously the extended periods needed for identification of viral etiologies; which greatly depended upon virus isolation techniques, meant that most viral diagnoses were of epidemiological value only [1–4].

Advances made in diagnostic techniques over the past decade have significantly improved the accuracy and timeliness of a viral diagnosis, which in turn can aid in patient management, disease control and positively impact the disease outcome [1–4]. Since the development of antiviral drugs and treatment options available for viral infections, clinicians are encouraged to seek viral laboratory diagnosis that can provide clinically useful information in diagnosis and management of patients. This required the focus of laboratories to shift to providing better, faster diagnosis, which has driven the development of new approaches to monitor viral infections and to support antiviral treatment through: quantitative viral loads, antiviral susceptibility testing, viral genotyping and, point-of-care testing. Despite the massive impact that molecular diagnostics has had on viral diagnosis, significant strides have been made in antigen detection and serological tests, in development of "rapid tests", for the direct detection of viral antigen in clinical specimens and detection of antibodies in convalescent or chronic infections. Laboratory controlled molecular and serological tests continue to have the advantage of superior sensitivity, specificity and differential diagnostic options in a controlled environment [1–4]. With the increase in sensitivity, specificity and diversity of virological diagnostic assays available, the clinician should work in collaboration with the virology laboratory to maximize the diagnostic potential of an appropriate clinical specimen. Understanding the relevance of the diagnostic test requested for specific viruses, at different ages and interpretation of a positive test, remains key in the clinical management of a patient [1–4].

The aim of this chapter is to provide an overview of diagnostic methodologies available for viral diagnosis rather than extensive technical details of each of the assays. It aims to provide an overview of options available for the clinician, from common assays to recent developments; the rationale for using each and how they could be successfully employed for better clinical management of patients.

6.2   Collecting and Sending Clinical Samples
to the Laboratory

The most important factor influencing the accuracy of viral diagnostic results is the specimen. Whichever method is used in the laboratory, the results are largely dependent upon the right specimen type, taken at the right time and stored and transported correctly [5].

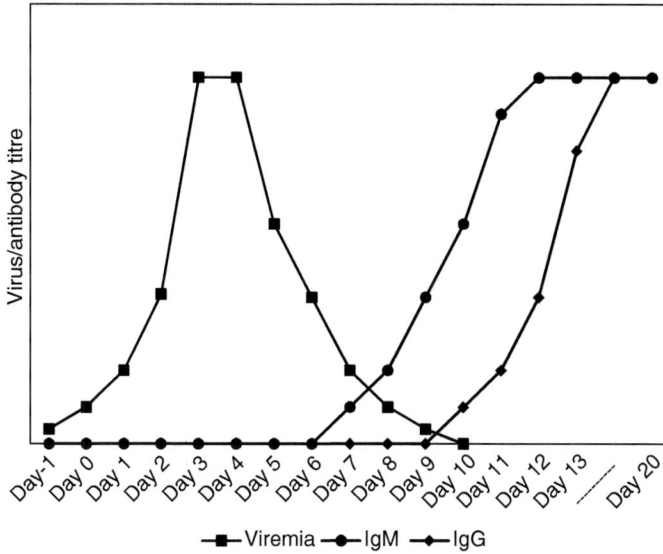

Fig. 6.1 Diagram indicating a typical acute virus infection showing the period that virus, IgM and IgG antibody can be detected (compiled from [4, 6, 8])

An understanding of the pathogenesis and epidemiology of the virus involved will help to identify the correct test and specimen type to collect. For acute viruses it is crucial to take the time since infection into consideration and whether the virus circulates commonly in the population and causes reinfections. Figure 6.1 depicts the typical period that acute viruses can be detected in blood and the time before IgM becomes visible and later IgG [4, 6]. For many acute viruses the viremia varies but may be relatively short and virus can only be detected in the first 10 days from the time that clinical symptoms became apparent, either in the blood or urine if it causes a systemic infection, such as arboviruses or measles; in the stool for enteric viruses such as rotavirus or poliovirus; in the site of infection such as the respiratory tract for respiratory viruses; or central nervous system for neuroinvasive infections. During this time virus specific tests such as virus isolation, antigen tests or molecular tests such as reverse transcription (RT) polymerase chain reaction (PCR) are appropriate. For viruses that are less common in the environment such as the arboviruses and childhood diseases prior to vaccination, IgM antibody tests can be requested after 7–10 days, but may not be detected in early specimens and are not appropriate for common viruses such as the respiratory viruses that may cause frequent reinfections. IgG antibody would only be used to diagnose acute infections if a paired serum is available 10–14 days apart and is not used for common viruses that may cause frequent reinfections. IgG testing may also be used to determine immune status following vaccines. Maternal antibody will interfere with IgG testing the first 4–6 months of life and therefore, for chronic diseases such as HIV transmission in children from HIV-infected mothers, DNA PCR testing is more suitable [4, 7, 8].

6.2.1 Type of Specimen

Specimens to be used for virus isolation and RTPCR should be kept below 4 °C (39.2 °F) and reach the laboratory within 72 hours to keep RNA intact. For enveloped single stranded RNA viruses such as RSV the success rate declines from 48 hours and all effort should be made to keep the specimen on ice from the time it is collected until it reaches the laboratory.

Blood: the usual required volume is between 2 and 10 ml depending on the patient's age, with the appropriate tube determined by the test required, and the appropriate blood component for the test (whole blood, plasma, serum). Anticoagulants such as heparin may inhibit PCR and EDTA tubes are preferred for molecular testing. For serology, clotted blood may be collected in SST (serum separation tubes) tubes that allow separation of red blood cells and serum through centrifugation. Virus isolation may be preferred from whole blood (EDTA) or serum (clotted blood) depending on the virus. *Swabs:* Swabs with a Dacron or rayon tip are preferred to ensure cells are collected and should be placed in viral transport medium that will preserve labile viruses for viral isolation and RTPCR. Washes or aspirates such as nasopharyngeal aspirates and other fluids such as saliva and urine should be placed in viral transport medium, although CSF is usually preferred undiluted. *Stool:* Obtain at least 4 g of stool and place in a sterile container. *Tissue*: Place in a sterile container with small amount of viral transport medium, for viral diagnosis. Specimens other than clotted blood must be kept at 4 °C (39.2 °F) and transported on ice to retain viability of the viruses and keep nucleic acids intact [7]. Table 6.1 summarizes the type of specimen and relevant tests available for viruses associated with different syndromes that may affect children (and adults).

6.3 Methods Used in Diagnostic Virology

6.3.1 Electron Microscopy

Although this is one of the oldest techniques it is not routinely used in diagnostic laboratories anymore. Electron microscopy (EM) is the only method available for directly visualizing the virus, and therefore has many applications beyond being purely diagnostic. The visualization of viruses with EM involves negative staining of the clinical specimen. Negative staining of the clinical sample is a relatively straightforward; inexpensive technique that would represent a "catch all" method of viral identification. EM could be particularly useful in identifying fastidious [87] or non-cultivable [88–90] virus in specimens, providing they have a high virus concentration with a sensitivity limit of approximately 10^6 viral particles per milliliter of specimen, making a negative result difficult to interpret [2]. While the sensitivity could be increased by ultracentrifugation or antibody-induced clumping, a further limitation is the lack of specificity, as EM can only identify up to the family level whereafter, other methods would have to be applied for a specific diagnosis [3].

Table 6.1 Specimen information for diagnostic virology; compiled from [1, 4, 9]

General syndrome	Agent	Specimen required	Diagnostic test options		Diagnostic ELISA		PCR: Commercially or in-house assays	Comments
			Commercial Rapid antigen detection test (RADT) and IFA available	Virus Isolation period in days	IgM	IgG		
Respiratory (pharyngitis, croup, bronchitis, pneumonia)	Adenovirus	Combination NP/OP swab; NPA, BAL	Yes	21 or Shell-vial	N/A	N/A	Yes [10, 11]	Virus has no specific seasonality and is detected all year round [10, 12, 13]. Shell vial tests followed by antibody staining allows detection in cells after 48–96 hours
	Coronavirus	Combination NP/OP swab; NPA, BAL	No	RL	N/A	N/A	Yes [14–16]	This includes emerging viruses SARS-CoV and MERS-CoV [17]
	Cytomegalovirus	Combination NP/OP swab; NPA, BAL, Blood	Yes	28	Yes	N/A	Yes [18]	CMV pneumonia in severely immunosuppressed HIV-positive patients and congenital infections [18]
	Enterovirus	Combination NP/OP swab; NPA, BAL	No	14	N/A	N/A	Yes [10, 12, 13, 19–21]	Virus has no specific seasonality and is detected all year round [10, 12, 13]

(continued)

Table 6.1 (continued)

General syndrome	Agent	Specimen required	Diagnostic test options					Comments
			Commercial Rapid antigen detection test (RADT) and IFA available	Virus Isolation period in days	Diagnostic ELISA		PCR: Commercially or in-house assays	
					IgM	IgG		
	Herpes simplex virus (HSV)	Combination NP/OP swab; NPA, BAL	No	1–7	Yes	N/A	Yes	HSV-1 has been associated with severe acute respiratory disease in severely immunocompromised patients [22]
	Human metapneumovirus	Combination NP/OP swab; NPA, BAL	No	2–21	N/A	N/A	Yes [10, 12, 15, 16, 23–25]	Virus not routinely cultured and IFA not available. Seasonality similar to that of RSV [10, 12, 13, 25].
	Influenza virus	Combination NP/OP swab; NPA, BAL	Yes	2–14	N/A	N/A	Yes [10, 24, 26, 27]	Antigen detection 40–90% sensitive. Seasonal, test for Influenza A and B. Subtyping for seasonal subtypes (H1N1pdm09/H3N2) by specialist laboratories
	Parainfluenza virus (PIV)	Combination NP/OP swab; NPA, BAL	Yes	2–14	N/A	N/A	Yes	Of all the PIVs, PIV3 is the main contributor to respiratory disease with a seasonality in the spring and summer months, PIV1, 2 and 4 less common [12, 13, 28].

	Respiratory syncytial virus	Combination NP/OP Swab in VTM; NPA, BAL	Yes	2–21	N/A	N/A	Yes [10, 15, 27, 29–34]	Rapid antigen detection variable specificity; 90% sensitive [35, 36]; Strong seasonal trends in autumn–winter months in South Africa; winter in temperate climates in Northern hemisphere, rainy season in tropics [10, 12, 13].
	Rhinovirus	Nasopharyngeal (NP) aspirate (NPA); NP wash; or NP swab, Oropharyngeal swab (OP), Combination NP/OP swab	No	2–7	No	No	Yes [19, 37, 38]	Too many strains to type serologically. Virus has no specific seasonality and is detected all year round [10, 12, 13].
Exanthem	Maculopapular	Arboviruses	CSF if neurological; plasma/serum for febrile, VHF	No	RL	Yes	Rise in antibody levels: paired sera 10–14 days apart	Yes [39–44]	Viremia short, RTPCR in first 10 days only making acute serum for IgM serology important. Neurological cases detected in CSF. Specific to geographic region, mostly dengue, Zika. Cross reactivity for flaviviruses complicate confirmation by serum neutralization assays required

(continued)

Table 6.1 (continued)

General syndrome	Agent	Specimen required	Commercial Rapid antigen detection test (RADT) and IFA available	Virus Isolation period in days	Diagnostic ELISA IgM	Diagnostic ELISA IgG	PCR: Commercially or in-house assays	Comments
	Enterovirus	CSF when clinically relevant; OP or rectal swab	No	14	N/A	N/A	Yes [10, 12, 13, 19–21]	
	Human herpes virus 6 and 7	Serum	No	RL	Yes	No	Yes	Roseola agent
	Measles virus	Serum, urine, respiratory secreta or CSF depending on syndrome	Yes	RL	Yes	Yes	Yes	Difficult to grow; RTPCR during acute, IgM serology later in disease for diagnostic purposes with paired sera [1, 9]
	Parvovirus B19	Serum	No	No	Yes	ND	Yes [45–50]	Erythema infectious agent; IgM serology is often diagnostic, but may be positive for a prolonged period [1]
	Rubella virus	CSF when clinically relevant, serum, urine	No	>10	Yes	Yes	Yes	Recommended that paired sera be tested simultaneously for diagnostic purposes [1, 9]
Vesicular	Herpes simplex virus	CSF when clinically relevant; Vesicle fluid, serum, EDTA	Yes	21	Yes	Yes	Yes [51, 52]	Serology rarely used for herpes simplex; IgM antibody used in selected cases [1]. Vesicle scrapings for direct IFA test [9]

	Specimen						Comment
CNS (Aseptic meningitis and encephalitis)							
Varicella-zoster virus	CSF when clinically relevant; Vesicle fluid, serum, EDTA	Yes	21	Yes	Yes	Yes [52]	Vesicle scrapings for direct IFA test
Arboviruses	CSF when clinically relevant and serum	No	No	Yes	Yes	Yes	Viremia brief, negative molecular test should be followed up with IgM serology with acute serum
Dengue	CSF when clinically relevant, serum	Yes	RL	Yes [53]	Yes	Yes [54–57]	According to CDC; 80% seropositive at 6 days. Viremia brief, negative molecular test should be followed up with IgM serology with acute serum
Cytomegalovirus	CSF when clinically relevant	No	RL	Yes	Yes	Yes [51]	Immunocompromised patients; newborns
Enterovirus	CSF when clinically relevant	No	3	No	No	Yes [58–60]	Most common viral cause of meningoencephalitis
Epstein-Barr virus	CSF	No	No	Yes	Yes	Yes [51]	Neurological cases confirmed on CSF, positive on serum need to be interpreted with caution due to reactivation
Hantavirus	Serum, CSF when clinically relevant	No	RL	Yes	ND	RL	Diagnosis by presence of IgM antibody, Dependant on geographic location (Americas/Europe)

(continued)

Table 6.1 (continued)

General syndrome	Agent	Specimen required	Diagnostic test options					Comments
			Commercial Rapid antigen detection test (RADT) and IFA available	Virus Isolation period in days	Diagnostic ELISA		PCR: Commercially or in-house assays	
					IgM	IgG		
	Herpes simplex virus	CSF when clinically relevant, serum	Yes	21	Yes	Yes	Yes [51, 61]	Culture of CSF has very low sensitivity; RTPCR on CSF most reliable
	Measles virus	CSF when clinically relevant, serum	Yes	21	Yes	Yes	Yes	Culture of virus usually very successful. IgM serology diagnostic and RTPCR
	Mumps virus	CSF when clinically relevant, urine	No	21	Yes [62]	Yes	Yes [63, 64]	RTPCR/IgM ELISA may allow diagnosis
	Rabies virus	Saliva, nuchal skin biopsy, CSF (pre-mortem); brain biopsy (postmortem)	Not WHO endorsed	2–3	N/A	Neutralizing antibodies on CSF only	(WHO), no commercial tests	Antigen detection (FAT) on tissue (brain); RTPCR tissue; CSF, saliva; Serology only on unvaccinated individual
	Varicella-zoster virus	CSF when clinically relevant, vesicular/skin swab	Yes	3–21 days [1]	Yes	Yes	Yes [51, 52]	

	West Nile virus	CSF when clinically relevant, EDTA blood; serum	Yes	RL	Yes	Yes	Yes [39, 65–67]	Viremia brief, RTPCR on CSF or EDTA blood; if negative follow up with IgM serology with acute serum. Cross reaction of flaviviruses require confirmation by serum neutralization assays
Infectious mononucleosis	Epstein-Barr virus	Serum	No	No	Yes	Yes	Yes [51]	PCR on blood most common but correlate with clinical presentation. Serology for infection status only
Hepatitis	Hepatitis A virus	Serum	No	No	Yes	ND	Yes [68–70]	Diagnosis by presence of IgM antibody/PCR
	Hepatitis B virus	Serum	Yes	No	Yes	ND	Yes	IgM to surface or core antigen confirms diagnosis; HBV viral load by PCR used for disease progression and clinical management
	Hepatitis C virus	Serum	No	No	Yes	ND	Yes	Serology and PCR confirms infection; viral load by PCR used for management
	Hepatitis D virus	Serum	No	No	Yes	No	No	Virus not routinely cultured
	Hepatitis E virus	Serum, stool	No	No	Yes [71]	Yes	Yes [72, 73]	Virus not routinely cultured

(continued)

Table 6.1 (continued)

General syndrome	Agent	Specimen required	Diagnostic test options						Comments
			Commercial Rapid antigen detection test (RADT) and IFA available	Virus Isolation period in days	Diagnostic ELISA			PCR: Commercially or in-house assays	
					IgM	IgG			
Immunodeficiency virus	HIV1/2	EDTA and serum	Yes	15	Yes 5–28	ND?		Yes	Antibody confirms infection in adults; DNA PCR for children <18 months due to passively acquired maternal antibody. Viral load by realtime PCR used for disease progression and clinical management
Gastroenteritis	Adenovirus	Rectal swab, stool	Yes	10	No	No		Yes [74–77]	Adenovirus 40 and 41 implicated in pediatric gastroenteritis
	Astrovirus	Stool	Yes	RL	No	No		Yes	Diagnosis by electron microscopy
	Norovirus or Norwalk	Stool	No	RL	No	No		Yes [78–82]	One of the major causes of acute gastroenteritis in communities or cruise ships
	Rotavirus	Stool	Yes	RL	No	No		Yes [76, 77, 83, 84]	Rapid assay are usually reliable, seasonal

Congenital infections						
Cytomegalovirus	EDTA, serum, amniotic fluid, cord blood	Yes	2	Yes	N/A	Yes [51]	Presence of CMV IgM in cord blood is indicative of congenital infections; PCR and culture on dried blood spots, blood, saliva or urine in newborn < 3 weeks; retrospective, should be done on stored dried blood spots taken after birth [8]
Enterovirus	CSF, serum, cord blood	No	3	Yes	N/A	Yes [58–60]	Cord blood most appropriate specimen in congenital infections
Herpes simplex virus	CSF, dermal, vesicle swab, tissue biopsy, amniotic fluid, cord blood	Yes	1	Yes	N/A	Yes [51]	Presence to HSV IgM in cord blood indicative of congenital infections
Parvovirus B19	Synovial fluid, amniotic fluid, plasma	No	No	Yes	ND	Yes [45–50]	
Rubella virus	CSF, serum	No	>10	Yes	N/A	Yes	IgM for Rubella should be assayed using serum from infants up to 6 months of age IgG; could be false positive due to maternal antibodies

(continued)

Table 6.1 (continued)

General syndrome	Agent	Specimen required	Diagnostic test options					Comments
			Commercial Rapid antigen detection test (RADT) and IFA available	Virus Isolation period in days	Diagnostic ELISA		PCR: Commercially or in-house assays	
					IgM	IgG		
	Zika virus	CSF, EDTA blood, serum. Pregnant mother/infant	No	RL	Yes	N/A	Yes	RTPCR recommended tests; IgM commercial tests need to be confirmed by PRNT by a reference laboratory for arboviruses due to cross reaction with dengue and other flaviviruses, depending on geographic relevance or travel history (CDC testing regiment [85, 86])

CSF cerebrospinal fluid, *ELISA* enzyme-linked immunosorbent assay, *PCR* polymerase chain reaction, *EDTA* ethylenediaminetetraacetic acid (purple top tube), *NP* nasopharyngeal, *OP* oropharyngeal, *NPA* nasopharyngeal aspirate, *BAL* bronchoalveolar lavage, *IFA* immunofluorescent antibody, *RL* indicates specialist or research laboratory only, *ND* not done, *N/A* Not applicable

Although the major advantage of EM is the speed with which a result could be obtained (30 min), the high cost of the instrument and specialized training and expertize needed, coupled with the lack of sensitivity and specificity, does not make this a viable option for routine diagnostics [4, 8].

6.3.2 Histology/Cytology

Direct microscopy of stained histology or cytology specimens may, in some instances, give the first indication of viral involvement that involves cellular changes. For viruses such as CMV, VZV, HPV, BK and B19, specific cytological changes can be confirmed through staining for specific antigen or genome sequences, using antibody or nucleic acid probes. Specific PCR amplification techniques may outperform these techniques in sensitivity, although detection of antigen in tissue is highly specific [4, 8].

6.3.3 Virus Isolation

Viral tissue culture was traditionally the "gold standard" used for diagnosing virus infections [91]. However, in the last 10 years molecular techniques have become routine. Virus isolation needs to remain an important part of viral diagnostics in order to maintain a source for analyzing, not only genotypic changes, but also phenotypic changes in virus populations for vaccine relevance and epidemiology. This allows identification of changes in antigenicity, pathogenicity and viral characteristics to update vaccines, such as the influenza vaccine, to match circulating strains [92]. Quality of the specimen, the time that it takes to reach the laboratory and transport under cold chain will determine the success of virus isolation. Detection of viruses in cell culture requires a considerable expertise and is performed by microscope examination, looking for degenerative morphological changes in the cell monolayer. This is called the cytopathic effect (CPE). Not all viruses grow in all cell types or produce CPE and further antigen or nucleic detection methods are required to correctly identify the specific virus involved. Clinical specimens are usually inoculated onto several cell lines to provide an optimum environment for a range of viruses (Table 6.2) [93].

An adaption of traditional viral culture formats has been developed, which allows for more rapid detection of viruses, especially for viruses which are known to grow slowly in conventional cell culture. This is achieved by inoculating the specimen onto a microscope slide and centrifugation of the culture to enhance the infection rate (Shell vial assays). The enhanced detection rate may result from better contact between cells in the specimen and the cell culture, thus allowing for earlier and more extensive infection of the cell lines, as well as through the use of fluorescent-labeled (e.g. FITC) monoclonal antibodies directed to the viral antigen [93]. Nevertheless, most culture methods lack sensitivity and specificity relative to PCR. It remains, however, a catchall method of choice if the virus in question can be cultured [92].

Table 6.2 List of viruses commonly isolated in clinical laboratories; compiled from [4]

Virus	Rate of growth [1, 94][a]	Type of CPE that can be detected [94]	Most permissive cell line[b]
RNA viruses			
Enteroviruses	2–8	Retractile angular or tear-shaped cells	PMK
Rhinoviruses	4–10	Retractile rounding of cells	HDF
Influenza viruses	2–14	Swollen vacuolated cells	PMK
RSV	2–21	Syncytia seen only in Hep-2 cells	HEp-2
DNA viruses			
Adenoviruses	1–21	Aggregation and rounding of cells in grape-like structures	HEp-2
HSV	1–7	Retractile rounded cells	A549
VZV	5–10	Foci of enlarged cells	HDF
CMV	5–28	Small foci of enlarged cells	HDF

[a]Time in days needed for CPE to develop, depending on the initial viral load in the sample, the higher the viral load the quicker CPE will be detected
[b]*PMK* Primary monkey kidney, *HDF* human dermal fibroblasts, *HEp-2* human epithelial type-2, *A549* adenocarcinoma human alveolar basal epithelial cells

6.3.4 Nucleic Acid Detection Methods

Viruses can be detected directly in clinical samples using highly specific nucleic acid primers and probes that are complementary in sequence to RNA viruses, using RT-PCR or for DNA viruses, directly by PCR. Over the past 10 years, nucleic acid amplification tests have been developed for the major viruses of public health concern and have become the new benchmark for viral diagnoses. The published sensitivities and specificities are usually nearly 100% when compared with cell culture or antigen assays [92, 95–97]. In fact studies that have compared molecular assays, with tissue culture assays, have demonstrated significantly increased sensitivity, of up to 30% [92, 95, 96, 98, 99].

The development of real-time PCR, that incorporates the use of specific florescent labeled probes, has created the ability to monitor the DNA amplification process as it happens, or in "real-time" on a dedicated instrument that is capable of collecting the fluorescent data from every PCR cycle. The accumulation of the measured fluorescence at the end of every PCR cycle is plotted and displayed as a sigmoidal curve and when the data is analyzed a cycle threshold (Ct) value is assigned to each target's amplification when it is first detected. The Ct is the point at which the amplicons' fluorescence exceeds that of the background and this is indirectly proportionate to the initial concentration of the target DNA in the sample i.e. the higher the concentration in the initial sample the lower the Ct value will be [100–102].

Comparative studies have revealed that the detection of respiratory viruses using real-time reverse transcriptase polymerase chain reaction (rRT-PCR) assays is substantially more sensitive than using conventional methods such as viral culture and immunofluorescence assays (IFA) [92, 103, 104]. Furthermore, compared to conventional PCR and other real-time methods, multiplex rRT-PCR has the advantage of permitting simultaneous amplification of several viruses in a single reaction [16, 103, 104]. This facilitates cost-effective diagnosis, enabling the detection of multiple viruses in a single clinical specimen. Amplification of several viruses together may however sacrifice sensitivity of individual assays and much effort has gone into identifying ways of increasing sensitivity.

6.3.4.1 Multiplex PCR Assays

Multiplex PCR assays are now frequently used to detect the presence of a range of viruses involved in specific syndromes such as respiratory infections e.g. influenza virus (INF) A and B; [20, 95, 105–107], parainfluenza viruses (PIV) types 1, 2, 3 and 4 [19, 20, 95, 105, 107, 108]; human respiratory syncytial virus (RSV) [20, 105–108], human metapneumovirus (hMPV) [20, 105, 107], human rhinoviruses (RV) [19, 20, 105, 107], human coronaviruses (hCoV-229E, hCoV-OC43) and Severe Acute Respiratory Syndrome coronavirus (SARS-CoV) [20, 107], human enteroviruses (EV) [19, 20] and adenoviruses (AdV) [108]. Disadvantages, include higher start-up costs, higher reagent costs, and extensive and specific training for specialist laboratories and specialized equipment to run them [92].

The diversity, fastidious nature, short viremia periods of some pathogens that may cause the infection and lack of available diagnostic tests, can severely hamper the ability to identify etiologies to different clinical syndromes [81]. Several multiplex platforms have been developed either as a set of duplex assays, in-house and commercial [20, 104, 109] and multiplex systems that require specialized equipment to read and are now available for a range of syndromes [20, 104, 109]. The limiting factor has been sensitivity and most assays require PCR amplification before detecting products through a number of platforms. These include Mass-Tag PCR [110], microarray platforms [111], macro arrays [39], microbead based methods [20] and Taq-Man array cards [112–115]. Currently TaqMan array cards are increasing in popularity due to the ease with which these can be adapted for specific purposes. These assays have application in diagnosis of single cases or as part of epidemiological studies to describe the etiologies of specific syndromes. TaqMan array card (TAC; Life Technologies, Foster City, CA, USA) assays have been developed and used with success for several syndromes such as respiratory disease [63], enteric disease [70] and neonatal sepsis [81]. Once developed, TaqMan array cards are stable at 4 °C (39.2 °F) for 2 years and can be shipped at ambient temperature [116]. The TAC assay is a 384-well microfluidic array which consists of identical arrays in eight individual microfluidic channels, each of which can be loaded with nucleic acid extract from a clinical specimen or positive control [63, 70, 81, 116, 117]. The individual channels consist of 48 wells, each of which contains singleplex qPCR reactions targeting a different pathogen. Thus eight specimens are assayed per TAC card, and can simultaneously detect up to 48 pathogens per specimen. This makes the TAC assay popular for the following reasons: (1) minimal specimen volume required; (2) reduction in cross contamination of specimens due to the closed system format; (3) the ability to tailor the panel of pathogens detected as required; (4) proven efficacy of this technology in pathogen detection for similar studies; (5) and simple to use format [63, 70, 81, 116, 117].

6.3.4.2 Future Trends
Multiplex methods are becoming more common in routine diagnostic laboratories. However, most of the large scale methods described above are predominantly research based and used in epidemiological studies or in specialist laboratories, rather than routinely. Next generation sequencing methods, that make use of deep sequencing of all nucleic acids present in a sample, are currently mostly used for pathogen discovery or in specialist laboratories to detect outbreaks. In general amplification steps are still needed before this can be used on clinical specimens. In addition these techniques are too expensive to run on a large scale, in routine diagnostic laboratories. Even though these techniques are becoming more affordable they generate significant amounts of data that require both trained bioinformaticians to interpret the outputs and large computational systems. Nevertheless, development of automated systems for identifying viruses directly in clinical specimens, may in future make these techniques more accessible for routine diagnostics [118–120].

6.3.5 Application of Molecular Virology Diagnostics in Clinical Management of Patients

6.3.5.1 Qualitative vs. Quantitative PCR
The increased sensitivity that the development of molecular assays have highlighted, is that while there are advantages in identifying new viruses associated with disease, such as the respiratory viruses described in the last decade (including hMPV [23], hCoV NL63 [14] and HKU1) [121], it has also revealed flaws that make interpreting such a positive PCR result problematic. Recent literature has shown that specifically in the case of RV [27, 37, 38] that the virus was detected in asymptomatic as well as symptomatic patients. Due to the increased sensitivity of molecular assays it is possible to detect the presence of a virus at a low genome copy number, which may represent the pre- or post-syndromic phase of a viral infection, redefining the nature of viral disease and the clinical interpretation thereof [27, 38]. Interpretation of qualitative and quantitative PCR results as well as the application of the appropriate choice would require a close liaison with the virological laboratory.

Qualitative detection in specimens that are normally virus free: A good example of this is the diagnosis of viral or aseptic encephalitis, in which testing CSF for HSV, CMV, VZV or enteroviruses are diagnostic [122]. Qualitative PCR offers significant advantages in terms of speed, especially with the development of the point-of-care testing. Early diagnosis and treatment of CNS infections has been proven to improve the prognosis [117] and reduce unnecessary treatment and hospitalization [123]. Viruses that only exhibit low-levels of virus shedding in the absence of symptoms such as viral gastroenteritis, caused by rotavirus or norovirus could be detected in stool samples [81, 82].

Quantitative viral loads: Assays that can quantify the amount of virus in infected patients have proven to be the most valuable tool in the management of chronic viral infections. For many persistent viral infections, with transient low-level viremia, the onset of symptoms is associated with a spike in viral replication and thus a higher viral load, allowing the prediction of disease onset [8]. This allows for better clinical management of the patient, as the clinician can monitor the progression of the disease, the success of treatment, the emergence of drug resistance and understanding the pathogenesis of a particular virus of which HIV-1 and 2 [124, 125], CMV [116, 126], EBV [127], HBV [128, 129] and HCV [124] are but a few.

6.3.5.2 Antiviral Resistance

As the availability of antiviral drugs increases, more emphasis is placed on assays to determine the causes of treatment failure, of which antiviral resistance is one possible outcome [8]. The emergence of antiviral resistance has been documented for virtually all antiviral compounds, with the specific viral mutations associated with resistance becoming better understood [8, 125, 130–132]. Laboratory assays to determine drug resistance fall into two major categories:

Phenotypic assays: Phenotypic assays have largely been replaced by molecular based genotypic assays, however, they remain the gold standard for determining drug efficacy and susceptibility, as the concentration of the drug required to inhibit viral replication can be calculated. HIV is the best example [130, 131, 133, 134]. Phenotypic assays have the added advantage of giving a complete overview of all mutations observed. However, it is an expensive and laborious technique, of which the success will greatly depend on the level of training of staff and whether or not the specific virus culture-adapted strains are available [8].

Genotypic assays: While the development of RT-PCR as a genotypic assay, focusing on specific areas on the virus genome, have the added advantage of being rapid, relatively inexpensive and semi-quantitative (single point mutation assays, and allelic discrimination assays), it is difficult to interpret a single point mutation without all the required information that a phenotypic assay would provide [8]. The development of new automated sequencing methods have enabled the study of the genetic basis of drug resistance and made the assessment of virus isolates, with reduced drug susceptibility, more accessible [8]. The use of sequence based methods for testing for antiviral resistance have also have become routine in viral diagnostics, especially for HIV [132, 135–137] and HBV [138–140]. The biggest drawback of this technique, other than the expense, is the downstream analysis of sequencing data that is generated. Sequencing editing and interpretation is required, and in the case of HIV, the identification of resistance is dependent upon the recognition of specific sequence patterns on the software system used [141].

6.3.6 Serology

Serological techniques can either be targeted at the antigen, during the acute phase of infection, or to virus specific antibody later in infection. While virus specific antigen may only be detected in the first 10 days of acute infections, IgM antibody is detected within 7–14 days following infection and may remain for a month or more in the patient's blood after the infection was cleared. Therefore, IgM can be used to detect a recent infection. IgG antibody is detected after 10–14 days of infection and can be present for life. Sero-conversion to IgG is measured with paired sera taken during the acute and the convalescent phase, 10–14 days apart. A significant rise in antibody of fourfold increase is seen as a positive reaction and new infection, while a single IgG positive test may reflect infection any time in the past. In pediatric infections, maternal immunity needs to be taken into consideration in the first 6 months of life and may only be cleared by 18 months, and may therefore, interfere with serological diagnosis in infants. Assays for detection of IgM or IgG are usually qualitative, since the presence or absence of antibody is enough to make a diagnosis. However, when a rise in antibody has to be detected, the test needs to be quantitative in order to detect an increase in antibody from the first to the second specimen. Antibody titre is measured as the reciprocal of the highest serum dilution where a positive reaction can still be detected. For example, a titre of 32 indicates that positive antibody binding could be detected in serum diluted up to 1 in 32, but not beyond that. Serological techniques are easily automated and play an important part in routine diagnostic laboratories. They have an important role in diagnosing acute and chronic infections and are useful for development of rapid tests and should, in addition, complement molecular techniques where clinically relevant [8, 142].

6.3.6.1 Immunofluorescence Assay (IFA)

These assays allow for the rapid detection of antigen and can be applied directly on clinical samples such as nasopharyngeal aspirates or on tissue culture or tissue from biopsy specimens, such as brain tissue for rabies virus. IFA is quick and convenient for individual specimens but requires a skilled operator and is not as easy to scale up and is not as sensitive as molecular techniques for viral detection. It is, however, relatively cheap and a popular choice, for this reason, in identification of viral infections.

Direct IFA detects virus in infected specimens or tissue using commercially available antibodies labelled with a florescent marker, while indirect IFA detects antibody in the patient sera by binding to the antigen in virus infected tissue cultured cells. A secondary anti-human IgG or IgM antibody is then used to detect the patient's bound antibodies. Direct IFA is frequently used for respiratory virus antigen detection in respiratory secretions (RSV, influenza, PIV 1–3) while indirect IFA for IgG or IgM antibody detection is used to detect infections such as EBV or VZV, amongst others [4, 8, 142, 143].

6.3.6.2 Enzyme-Linked Immunoassay for Antibody Detection (ELISA)

ELISAs are the most commonly used antibody or antigen detection assays since they have a high throughput, are rapid, are easily automated and are objective since the output can be read using a spectrophotometer. ELISA works on the principle of detecting antibody in patient sera through a reaction where antigen is bound to the surface of a micro-titre plate, the patient serum added to bind to the antigen and any bound antibody is then detected through addition of a secondary anti-human antibody coupled to an enzyme. Addition of a substrate to the enzyme linked antigen antibody complex results in a colour change which will induce a positive reaction. The assay can be adjusted for IgG or IgM through addition of the anti-human antibody. Antigen detection ELISA is performed by coating the solid phase with antibody to detect the antigen in the patient sera in order to reveal the detection antibody complex.

ELISA is frequently used for detection of IgG or IgM antibodies to rubella, measles, mumps, HIV, Hepatitis A and arboviruses such as West Nile virus, Zika virus, dengue or JEV. Although ELISA can have a very high sensitivity and specificity, some viral families may cross react and for exact identification of viruses such as the flaviviruses, neutralization assays are needed for confirmation [6, 142].

6.3.6.3 Neutralization Assays

Virus neutralization assays are highly specific assays testing for neutralizing antibodies and are also used to confirm results of other serological assays, such as ELISA, which are known to cross-react between different viruses of the same family e.g. the Flaviviruses (Zika virus, dengue and West Nile virus). It can also be used to determine if a vaccine would provide protection e.g. to detect antigenic drift in the neutralizing epitopes of the annual influenza vaccine. Antigen is mixed with dilutions of antibody and the inhibition of CPE observed through inoculation on a tissue monolayer. The inhibition effect can either be read through observation of CPE, or through overlay of agar which allows plaque formation for plaque reduction neutralization assays (PRNT). Micro-neutralization assays can also be read through ELISA methods, which help to automate the process and reduce the test run time before infected cells can be detected. These techniques are labour intensive and not routinely done by diagnostic laboratories, but rather by reference or specialist laboratories [6, 142, 144].

6.3.6.4 Other Serological Techniques

Several further formats of serological techniques exist that are used for different purposes. The **hemagglutination inhibition assay (HAI)** test detects antibodies to viruses that have a hemagglutinin antigen. These include rubella, Influenza and the flaviviruses. The test is still routinely used in reference laboratories to identify especially, cross reactive viruses such as the flaviviruses, before confirming specific viruses by neutralization assays or to investigate influenza antigenic variation relative to sera raised against the vaccine. Due to cross reactivity and requirement for fresh red blood cells, it is less commonly employed in routine laboratories. **The Western blot technique** is still used for confirmation of HIV and HCV, and it is based on the principle of transferring specific viral proteins separated on a gel or blotting paper, followed by binding to patient serum and detection with an anti-human enzyme labeled antibody and substrate. It is very specific but sensitivity may vary. Antigen detection methods allow for the development of rapid antigen or antibody tests and allow for bedside diagnosis. However, variable sensitivity and specificity determine the value of these tests [142].

6.3.6.5 Future Trends in Serology

Following the trend in molecular diagnostics, development of multiplex serological assays, that cover a range of viral antigens associated with specific syndromes, will significantly improve the diagnostic capacity of laboratories. Methods that use multiplex microsphere-based suspension immuno assays (SIAs) for the simultaneous detection of IgG antibodies against a range of viruses, enables development of syndrome or application specific tests. An example would be a B19, CMV and *T. gondii* combination SIAs multiplex for rapid antibody screening during pregnancy [145]. These assays bind a number of antigens through antibody to a microsphere. They are then incubated with the patient sera before being visualized with a labeled anti-human IgM antibody. Similar tests have been described for arbovirus screening in the Northern hemisphere [146]. Multiplex formats, based on protein arrays, have also been developed to detect a range of viruses. For these assays the antigens are fixed to a solid phase microchip slide and tested against patient anti-serum and fluorescent labeled anti-human antibodies used for detection. The position on the chip identifies the pathogen involved. These arrays may be based on peptides synthesized from pathogen sequence [147] recombinant proteins, [148] or inactivated virus antigens [149]. These techniques are not yet widely available in routine diagnostic laboratories but developed in specialist and research laboratories. However, the automation possibilities will very likely improve their accessible in the future.

6.3.7 Quality Assurance and Control

The ability of the laboratory to provide accurate diagnostic results is essential for effective clinical management of patients, solving of outbreaks and for responsible decision making [150]. Therefore monitoring ongoing quality assurance (QA) and improvement in all aspects of the laboratory is crucial. This involves the managing and monitoring of all services and processes related to releasing a diagnostic results [4, 150]. Processes that should be monitored relate to the pre-analytical phase *i.e.* specimen transport, collection and storage; the analytical phase *i.e.* testing and monitoring of the laboratory procedures and environment as well as the post analytical phase *i.e.* of result reporting and result interpretation [4, 150]. QA ensures annual assessment of staff competency, calibration and servicing of equipment as well as the quality of diagnostic tests. Clinical laboratories are strictly regulated by appointed agencies and are audited according to specific standards set forth by the International Organization for Standardization (ISO) and the International Electrotechnical Commission (IEC) [4, 150].

The primary quality control (QC) concern in a molecular laboratory is the specimen and nucleic acid quality or integrity, assay sensitivity and specificity, as well as the false positive tests because of PCR contamination. The RNA and DNA integrity can be insured through use of RNase and DNase free reagents and consumables, in addition to handling specimens on ice. While, PCR contamination can be avoided through physical organization of the laboratory and workflow, separating work areas and equipment; relevant PCR controls should be included in each run to ensure correct interpretation of the results [4, 8, 150]. The use of uracil N-glycosylase (UNG) in PCR reactions provides for chemical control for carry over contamination [151].

It is vital that when PCR diagnostics are undertaken, every effort is made to minimize contamination and that these assays are tested in a laboratory environment in which staff are well trained and competent for this type of work [8]. Using an accredited laboratory ensures the diagnostic findings are reliable.

6.4 Conclusion

Molecular and serological techniques should be used in a complimentary fashion for the diagnoses of virus infections. Knowledge of the stage of infection, viral pathogenesis and epidemiology help to make decisions regarding the correct test to choose for appropriate diagnosis. Advances in specificity and sensitivity and capacity to test for a range of viruses through multiple platforms make accurate diagnosis and rapid identification of circulating viruses for real-time clinical relevant data more feasible today. Virus isolation and collaboration with specialist laboratories make newer techniques for identification of emerging and re-emerging viruses possible. Quality of specimens, patient clinical history and presentation and close collaboration between the clinician, pathologist and laboratory remain key for useful diagnostic data for improved patient management.

References

1. Levin M, Asturias E, Weinberg A. Infections: viral & rickettsial, vol. 23e. New York, NY: McGraw-Hill; 2016.
2. Miller SE. Electron microscopy of viral infections. In: Lennette EH, Smith TF, editors. Laboratory diagnosis of viral infections. 3rd ed. New York, NY: Marcel Dekker, Inc.; 1999. p. 45–70.
3. Petric M, Szymanski M. Electron microscopy and immunoelectron microscopy. In: Specter S, Hodinka RL, Young SA, editors. Clinical virology manual. 3rd ed. Washington, DC: American Society of Microbiology Press; 2000. p. 54–65.
4. Specter S, Hodinka RL, Wiedbrauk DL, Young SA. Diagnosis of viral infections. In: Richman DD, Whitley RJ, Hayden FG, editors. Clinical virology, vol. 35. 2nd ed. Washington, DC: American Society for Microbiology Press; 2002. p. 243–72.
5. Nutting PA, Main DS, Fischer PM, Stull TM, Pontious M, Seifert Jr M, Boone DJ, Holcomb S. Toward optimal laboratory use. Problems in laboratory testing in primary care. JAMA. 1996;275:635–9.
6. Lambert AJ, Lanciotti RS. Laboratory diagnosis of arboviruses. Norfolk, UK: Caister Academic press; 2016.
7. Storch G. Diagnostic virology. In: Knipe DM, Howley PM, editors. Fields virology, vol. 5. 5th ed. Philadelphia, PA: Lippincott Williams & Wilkins; 2007. p. 565–604.
8. Jeffery K, Aarons E. Diagnostic approaches. In: Zuckerman AJ, Banatvala JE, Pattison JR, editors. Principles and practice of clinical virology. New York: Wiley; 2009. p. 1–27.
9. Grys TE, Smith TF. Specimen requirements: selection, collection, transport and processing. In: Steven S, Hodinka RL, Young SA, Wiedbrauk DL, editors. Clinical virology manual. 4th ed. Washington, DC: ASM Press; 2009. p. 18–35.

10. Pretorius MA, Madhi SA, Cohen C, Naidoo D, Groome M, Moyes J, Buys A, Walaza S, Dawood H, Chhagan M, Haffjee S, Kahn K, Puren A, Venter M. Respiratory viral coinfections identified by a 10-plex real-time reverse-transcription polymerase chain reaction assay in patients hospitalized with severe acute respiratory illness--South Africa, 2009-2010. J Infect Dis. 2012;206(Suppl 1)S159–65.
11. Heim A, Ebnet C, Harste G, Pring-Akerblom P. Rapid and quantitative detection of human adenovirus DNA by real-time PCR. J Med Virol. 2003;70:228–39.
12. Cohen C, Walaza S, Moyes J, Groome M, Tempia S, Pretorius M, Hellferscee O, Dawood H, Chhagan M, Naby F, Haffejee S, Variava E, Kahn K, Nzenze S, Tshangela A, von Gottberg A, Wolter N, Cohen AL, Kgokong B, Venter M, Madhi SA. Epidemiology of viral-associated acute lower respiratory tract infection among children <5 years of age in a high HIV prevalence setting, South Africa, 2009-2012. Pediatr Infect Dis J. 2015;34:66–72.
13. Cohen C, Walaza S, Moyes J, Groome M, Tempia S, Pretorius M, Hellferscee O, Dawood H, Haffejee S, Variava E, Kahn K, Tshangela A, von Gottberg A, Wolter N, Cohen AL, Kgokong B, Venter M, Madhi SA. Epidemiology of severe acute respiratory illness (SARI) among adults and children aged >/=5 years in a high HIV-prevalence setting, 2009-2012. PLoS One. 2015;10:e0117716.
14. van der Hoek L, Pyrc K, Jebbink MF, Vermeulen-Oost W, Berkhout RJ, Wolthers KC, Wertheim-van Dillen PM, Kaandorp J, Spaargaren J, Berkhout B. Identification of a new human coronavirus. Nat Med. 2004;10:368–73.
15. Lassauniere R, Kresfelder T, Venter M. A novel multiplex real-time RT-PCR assay with FRET hybridization probes for the detection and quantitation of 13 respiratory viruses. J Virol Methods. 2010;165:254–60.
16. Venter M, Lassauniere R, Kresfelder TL, Westerberg Y, Visser A. Contribution of common and recently described respiratory viruses to annual hospitalizations in children in South Africa. J Med Virol. 2011;83:1458–68.
17. Sisk JM, Frieman MB. Emerging coronaviruses: severe acute respiratory syndrome (SARS) and middle east respiratory syndrome (MERS), eLS 1–12. New York: John Wiley & Sons, Ltd; 2001.
18. Salomon N, Gomez T, Perlman DC, Laya L, Eber C, Mildvan D. Clinical features and outcome of HIV-related cytomegalovirus pneumonia. AIDS. 1997;11:319–24.
19. Coiras MT, Aguilar JC, Garcia ML, Casas I, Perez-Brena P. Simultaneous detection of fourteen respiratory viruses in clinical specimens by two multiplex reverse transcription nested-PCR assays. J Med Virol. 2004;72:484–95.
20. Mahony J, Chong S, Merante F, Yaghoubian S, Sinha T, Lisle C, Janeczko R. Development of a respiratory virus panel test for detection of twenty human respiratory viruses by use of multiplex PCR and a fluid microbead-based assay. J Clin Microbiol. 2007;45:2965–70.
21. Mahony JB. Detection of respiratory viruses by molecular methods. Clin Microbiol Rev. 2008;21:716–47.
22. Simoons-Smit AM, Kraan EM, Beishuizen A, van Schijndel RJS, Vandenbroucke-Grauls CM. Herpes simplex virus type 1 and respiratory disease in critically-ill patients: real pathogen or innocent bystander? Clin Microbiol Infect. 2006;12:1050–9.
23. van den Hoogen BG, de Jong JC, Groen J, Kuiken T, de Groot R, Fouchier RA, Osterhaus AD. A newly discovered human pneumovirus isolated from young children with respiratory tract disease. Nat Med. 2001;7:719–24.
24. Cohen C, Moyes J, Tempia S, Groome M, Walaza S, Pretorius M, Dawood H, Chhagan M, Haffejee S, Variava E, Kahn K, von Gottberg A, Wolter N, Cohen AL, Malope-Kgokong B, Venter M, Madhi SA. Mortality amongst patients with influenza-associated severe acute respiratory illness, South Africa, 2009-2013. PLoS One. 2015;10:e0118884.
25. Groome MJ, Moyes J, Cohen C, Walaza S, Tempia S, Pretorius M, Hellferscee O, Chhagan M, Haffejee S, Dawood H, Kahn K, Variava E, Cohen AL, Gottberg A, Wolter N, Venter M, Madhi SA. Human metapneumovirus-associated severe acute respiratory illness hospitalisation in HIV-infected and HIV-uninfected South African children and adults. J Clin Virol. 2015;69:125–32.

26. Cohen AL, Hellferscee O, Pretorius M, Treurnicht F, Walaza S, Madhi S, Groome M, Dawood H, Variava E, Kahn K, Wolter N, von Gottberg A, Tempia S, Venter M, Cohen C. Epidemiology of influenza virus types and subtypes in South Africa, 2009-2012. Emerg Infect Dis. 2014;20:1162–9.
27. Pretorius MA, Tempia S, Walaza S, Cohen AL, Moyes J, Variava E, Dawood H, Seleka M, Hellferscee O, Treurnicht F, Cohen C, Venter M. The role of influenza, RSV and other common respiratory viruses in severe acute respiratory infections and influenza-like illness in a population with a high HIV sero-prevalence, South Africa 2012–2015. J Clin Virol. 2016;75:21–6.
28. Cohen AL, Sahr PK, Treurnicht F, Walaza S, Groome MJ, Kahn K, Dawood H, Variava E, Tempia S, Pretorius M, Moyes J, Olorunju SA, Malope-Kgokong B, Kuonza L, Wolter N, von Gottberg A, Madhi SA, Venter M, Cohen C. Parainfluenza virus infection among human immunodeficiency virus (HIV)-infected and HIV-uninfected children and adults hospitalized for severe acute respiratory illness in South Africa, 2009-2014. Open Forum Infect Dis. 2015;2:ofv139.
29. Haynes AK, Manangan AP, Iwane MK, Sturm-Ramirez K, Homaira N, Brooks WA, Luby S, Rahman M, Klena JD, Zhang Y, Yu H, Zhan F, Dueger E, Mansour AM, Azazzy N, McCracken JP, Bryan JP, Lopez MR, Burton DC, Bigogo G, Breiman RF, Feikin DR, Njenga K, Montgomery J, Cohen AL, Moyes J, Pretorius M, Cohen C, Venter M, Chittaganpitch M, Thamthitiwat S, Sawatwong P, Baggett HC, Luber G, Gerber SI. Respiratory syncytial virus circulation in seven countries with Global Disease Detection Regional Centers. J Infect Dis. 2013;208(Suppl 3)S246–54.
30. Moyes J, Cohen C, Pretorius M, Groome M, von Gottberg A, Wolter N, Walaza S, Haffejee S, Chhagan M, Naby F, Cohen AL, Tempia S, Kahn K, Dawood H, Venter M, Madhi SA. Epidemiology of respiratory syncytial virus-associated acute lower respiratory tract infection hospitalizations among HIV-infected and HIV-uninfected South African children, 2010-2011. J Infect Dis. 2013;208(Suppl 3)S217–26.
31. Nair H, Brooks WA, Katz M, Roca A, Berkley JA, Madhi SA, Simmerman JM, Gordon A, Sato M, Howie S, Krishnan A, Ope M, Lindblade KA, Carosone-Link P, Lucero M, Ochieng W, Kamimoto L, Dueger E, Bhat N, Vong S, Theodoratou E, Chittaganpitch M, Chimah O, Balmaseda A, Buchy P, Harris E, Evans V, Katayose M, Gaur B, O'Callaghan-Gordo C, Goswami D, Arvelo W, Venter M, Briese T, Tokarz R, Widdowson M-A, Mounts AW, Breiman RF, Feikin DR, Klugman KP, Olsen SJ, Gessner BD, Wright PF, Rudan I, Broor S, Simões EAF, Campbell H. Global burden of respiratory infections due to seasonal influenza in young children: a systematic review and meta-analysis. Lancet. 2011;378:1917–30.
32. Pretorius MA, van Niekerk S, Tempia S, Moyes J, Cohen C, Madhi SA, Venter M. Replacement and positive evolution of subtype a and B respiratory syncytial virus G-protein genotypes from 1997-2012 in South Africa. J Infect Dis. 2013;208(Suppl 3)S227–37.
33. Tempia S, Walaza S, Viboud C, Cohen AL, Madhi SA, Venter M, McAnerney JM, Cohen C. Mortality associated with seasonal and pandemic influenza and respiratory syncytial virus among children <5 years of age in a high HIV prevalence setting—South Africa, 1998–2009. Clin Infect Dis. 2014;58:1241–9.
34. Venter M, Collinson M, Schoub BD. Molecular epidemiological analysis of community circulating respiratory syncytial virus in rural South Africa: comparison of viruses and genotypes responsible for different disease manifestations. J Med Virol. 2002;68:452–61.
35. Slinger R, Milk R, Gaboury I, Diaz-Mitoma F. Evaluation of the QuickLab RSV Test, a new rapid lateral-flow immunoassay for detection of respiratory syncytial virus antigen. J Clin Microbiol. 2004;42:3731–3.
36. Englund JA, Piedra PA, Jewell A, Patel K, Baxter BB, Whimbey E. Rapid diagnosis of respiratory syncytial virus infections in immunocompromised adults. J Clin Microbiol. 1996;34:1649–53.
37. Fry AM, Lu X, Olsen SJ, Chittaganpitch M, Sawatwong P, Chantra S, Baggett HC, Erdman D. Human rhinovirus infections in rural Thailand: epidemiological evidence for rhinovirus as both pathogen and bystander. PLoS One. 2011;6:e17780.

38. Pretorius MA, Tempia S, Treurnicht FK, Walaza S, Cohen AL, Moyes J, Hellferscee O, Variava E, Dawood H, Chhagan M, Haffjee S, Madhi SA, Cohen C, Venter M. Genetic diversity and molecular epidemiology of human rhinoviruses in South Africa. Influenza Other Respi Viruses. 2014;8:567–73.

39. Zaayman D, Human S, Venter M. A highly sensitive method for the detection and genotyping of West Nile virus by real-time PCR. J Virol Methods. 2009;157:155–60.

40. Garcia S, Crance JM, Billecocq A, Peinnequin A, Jouan A, Bouloy M, Garin D. Quantitative real-time PCR detection of rift valley fever virus and its application to evaluation of antiviral compounds. J Clin Microbiol. 2001;39:4456–61.

41. Moureau G, Temmam S, Gonzalez JP, Charrel RN, Grard G, de Lamballerie X. A real-time RT-PCR method for the universal detection and identification of flaviviruses. Vector Borne Zoonotic Dis. 2013;7(4)467–78.

42. Patel P, Landt O, Kaiser M, Faye O, Koppe T, Lass U, Sall A, Niedrig M. Development of one-step quantitative reverse transcription PCR for the rapid detection of flaviviruses. Virol J. 2013;10:58.

43. Van Eeden C, Zaayman D, Venter M. A sensitive nested real-time RT-PCR for the detection of Shuni virus. J Virol Methods. 2014;195:100–5.

44. Forrester NL, Palacios G, Tesh RB, Savji N, Guzman H, Sherman M, Weaver SC, Lipkin WI. Genome-scale phylogeny of the alphavirus genus suggests a marine origin. J Virol. 2012;86:2729–38.

45. Zaaijer HL, Koppelman MH, Farrington CP. Parvovirus B19 viraemia in Dutch blood donors. Epidemiol Infect. 2004;132:1161–6.

46. Koppelman MH, Cuypers HT, Emrich T, Zaaijer HL. Quantitative real-time detection of parvovirus B19 DNA in plasma. Transfusion. 2004;44:97–103.

47. Weimer T, Streichert S, Watson C, Groner A. High-titer screening PCR: a successful strategy for reducing the parvovirus B19 load in plasma pools for fractionation. Transfusion. 2001;41:1500–4.

48. Schmidt I, Blumel J, Seitz H, Willkommen H, Lower J. Parvovirus B19 DNA in plasma pools and plasma derivatives. Vox Sang. 2001;81:228–35.

49. Harder TC, Hufnagel M, Zahn K, Beutel K, Schmitt HJ, Ullmann U, Rautenberg P. New LightCycler PCR for rapid and sensitive quantification of parvovirus B19 DNA guides therapeutic decision-making in relapsing infections. J Clin Microbiol. 2001;39:4413–9.

50. Aberham C, Pendl C, Gross P, Zerlauth G, Gessner M. A quantitative, internally controlled real-time PCR assay for the detection of parvovirus B19 DNA. J Virol Methods. 2001;92:183–91.

51. Espy MJ, Uhl JR, Sloan LM, Buckwalter SP, Jones MF, Vetter EA, Yao JD, Wengenack NL, Rosenblatt JE, Cockerill 3rd FR, Smith TF. Real-time PCR in clinical microbiology: applications for routine laboratory testing. Clin Microbiol Rev. 2006;19:165–256.

52. O'Neill HJ, Wyatt DE, Coyle PV, McCaughey C, Mitchell F. Real-time nested multiplex PCR for the detection of Herpes simplex virus types 1 and 2 and Varicella zoster virus. J Med Virol. 2003;71:557–60.

53. Shu PY, Chen LK, Chang SF, Yueh YY, Chow L, Chien LJ, Chin C, Lin TH, Huang JH. Comparison of capture immunoglobulin M (IgM) and IgG enzyme-linked immunosorbent assay (ELISA) and nonstructural protein NS1 serotype-specific IgG ELISA for differentiation of primary and secondary dengue virus infections. Clin Diagn Lab Immunol. 2003;10:622–30.

54. Shu PY, Chang SF, Kuo YC, Yueh YY, Chien LJ, Sue CL, Lin TH, Huang JH. Development of group- and serotype-specific one-step SYBR green I-based real-time reverse transcription-PCR assay for dengue virus. J Clin Microbiol. 2003;41:2408–16.

55. Houng HS, Chung-Ming Chen R, Vaughn DW, Kanesa-thasan N. Development of a fluorogenic RT-PCR system for quantitative identification of dengue virus serotypes 1-4 using conserved and serotype-specific 3′ noncoding sequences. J Virol Methods. 2001;95:19–32.

56. Callahan JD, Wu SJ, Dion-Schultz A, Mangold BE, Peruski LF, Watts DM, Porter KR, Murphy GR, Suharyono W, King CC, Hayes CG, Temenak JJ. Development and evaluation

of serotype- and group-specific fluorogenic reverse transcriptase PCR (TaqMan) assays for dengue virus. J Clin Microbiol. 2001;39:4119–24.

57. Laue T, Emmerich P, Schmitz H. Detection of dengue virus RNA in patients after primary or secondary dengue infection by using the TaqMan automated amplification system. J Clin Microbiol. 1999;37:2543–7.

58. Monpoeho S, Coste-Burel M, Costa-Mattioli M, Besse B, Chomel JJ, Billaudel S, Ferre V. Application of a real-time polymerase chain reaction with internal positive control for detection and quantification of enterovirus in cerebrospinal fluid. Eur J Clin Microbiol Infect Dis. 2002;21:532–6.

59. Corless CE, Guiver M, Borrow R, Edwards-Jones V, Fox AJ, Kaczmarski EB, Mutton KJ. Development and evaluation of a 'real-time' RT-PCR for the detection of enterovirus and parechovirus RNA in CSF and throat swab samples. J Med Virol. 2002;67:555–62.

60. Verstrepen WA, Kuhn S, Kockx MM, Van De Vyvere ME, Mertens AH. Rapid detection of enterovirus RNA in cerebrospinal fluid specimens with a novel single-tube real-time reverse transcription-PCR assay. J Clin Microbiol. 2001;39:4093–6.

61. Aberle SW, Puchhammer-Stockl E. Diagnosis of herpesvirus infections of the central nervous system. J Clin Virol. 2002;25(Suppl 1)S79–85.

62. Krause CH, Molyneaux PJ, Ho-Yen DO, McIntyre P, Carman WF, Templeton KE. Comparison of mumps-IgM ELISAs in acute infection. J Clin Virol. 2007;38:153–6.

63. Krause CH, Eastick K, Ogilvie MM. Real-time PCR for mumps diagnosis on clinical specimens--comparison with results of conventional methods of virus detection and nested PCR. J Clin Virol. 2006;37:184–9.

64. Uchida K, Shinohara M, Shimada S, Segawa Y, Doi R, Gotoh A, Hondo R. Rapid and sensitive detection of mumps virus RNA directly from clinical samples by real-time PCR. J Med Virol. 2005;75:470–4.

65. Lanciotti R. Molecular amplification assays for the detection of flaviviruses. Adv Virus Res. 2003;61:67–99.

66. Lanciotti RS, Ebel GD, Deubel V, Kerst AJ, Murri S, Meyer R, Bowen M, McKinney N, Morrill WE, Crabtree MB, Kramer LD, Roehrig JT. Complete genome sequences and phylogenetic analysis of West Nile virus strains isolated from the United States, Europe, and the Middle East. Virology. 2002;298:96–105.

67. Zaayman D, Venter M. West Nile Virus neurologic disease in humans, South Africa, September 2008–May 2009. Emerg Infect Dis. 2012;18:2051–4.

68. Sanchez G, Populaire S, Butot S, Putallaz T, Joosten H. Detection and differentiation of human hepatitis A strains by commercial quantitative real-time RT-PCR tests. J Virol Methods. 2006;132:160–5.

69. Brooks HA, Gersberg RM, Dhar AK. Detection and quantification of hepatitis A virus in seawater via real-time RT-PCR. J Virol Methods. 2005;127:109–18.

70. Costa-Mattioli M, Di Napoli A, Ferre V, Billaudel S, Perez-Bercoff R, Cristina J. Genetic variability of hepatitis A virus. J Gen Virol. 2003;84:3191–201.

71. Mansuy JM, Peron JM, Bureau C, Alric L, Vinel JP, Izopet J. Immunologically silent autochthonous acute hepatitis E virus infection in France. J Clin Microbiol. 2004;42:912–3.

72. Orru G, Masia G, Orru G, Romano L, Piras V, Coppola RC. Detection and quantitation of hepatitis E virus in human faeces by real-time quantitative PCR. J Virol Methods. 2004;118:77–82.

73. Mansuy JM, Peron JM, Abravanel F, Poirson H, Dubois M, Miedouge M, Vischi F, Alric L, Vinel JP, Izopet J. Hepatitis E in the south west of France in individuals who have never visited an endemic area. J Med Virol. 2004;74:419–24.

74. Banerjee A, De P, Manna B, Chawla-Sarkar M. Molecular characterization of enteric adenovirus genotypes 40 and 41 identified in children with acute gastroenteritis in Kolkata, India during 2013–2014. J Med Virol. 2016;89:606–14.

75. Sanaei Dashti A, Ghahremani P, Hashempoor T, Karimi A. Molecular epidemiology of enteric adenovirus gastroenteritis in under-five-year-old children in Iran. Gastroenterol Res Pract. 2016;2016:2045697.
76. Ozsari T, Bora G, Kaya B, Yakut K. The prevalence of rotavirus and adenovirus in the childhood gastroenteritis. Jundishapur J Microbiol. 2016;9:e34867.
77. Liu L, Qian Y, Zhang Y, Zhao L, Jia L, Dong H. Epidemiological aspects of rotavirus and adenovirus in hospitalized children with diarrhea: a 5-year survey in Beijing. BMC Infect Dis. 2016;16:1–7.
78. Qiao N, Wang X-Y, Liu L. Temporal evolutionary dynamics of norovirus GII.4 variants in China between 2004 and 2015. PLoS One. 2016;11:e0163166.
79. Kabue JP, Meader E, Hunter PR, Potgieter N. Norovirus prevalence and estimated viral load in symptomatic and asymptomatic children from rural communities of Vhembe district, South Africa. J Clin Virol. 2016;84:12–8.
80. Brown JR, Shah D, Breuer J. Viral gastrointestinal infections and norovirus genotypes in a paediatric UK hospital, 2014–2015. J Clin Virol. 2016;84:1–6.
81. O'Neill HJ, McCaughey C, Coyle PV, Wyatt DE, Mitchell F. Clinical utility of nested multiplex RT-PCR for group F adenovirus, rotavirus and norwalk-like viruses in acute viral gastroenteritis in children and adults. J Clin Virol. 2002;25:335–43.
82. O'Neill HJ, McCaughey C, Wyatt DE, Mitchell F, Coyle PV. Gastroenteritis outbreaks associated with Norwalk-like viruses and their investigation by nested RT-PCR. BMC Microbiol. 2001;1:14.
83. Do LP, Kaneko M, Nakagomi T, Gauchan P, Agbemabiese CA, Dang AD, Nakagomi O. Molecular epidemiology of Rotavirus A, causing acute gastroenteritis hospitalisations among children in Nha Trang, Vietnam, 2007–2008: Identification of rare G9P[19] and G10P[14] strains. J Med Virol. 2016;89:621–31.
84. Wandera EA, Mohammad S, Komoto S, Maeno Y, Nyangao J, Ide T, Kathiiko C, Odoyo E, Tsuji T, Taniguchi K, Ichinose Y. Molecular epidemiology of rotavirus gastroenteritis in Central Kenya before vaccine introduction, 2009–2014. J Med Virol. 2016;89:809–17.
85. CDC. Guidance for U.S. laboratories testing for Zika virus infection. Atlanta, GA: Centres for Disease Control and Prevention; 2016. http://www.cdc.gov/zika/laboratories/lab-guidance.html. Accessed 10 Mar 2016
86. Petersen EE, Polen KND, Meaney-Delman D, Ellington SR, Oduyebo T, Cohn A, Oster AM, Russell K, Kawwass JF, Karwowski MP, Powers AM, Bertolli J, Brooks JT, Kissin D, Villanueva J, Muñoz-Jordan J, Kuehnert M, Olson CK, Honein MA, Rivera M, Jamieson DJ, Rasmussen SA. Update: interim guidance for health care providers caring for women of reproductive age with possible Zika virus exposure — United States. MMWR Morb Mortal Wkly Rep. 2016;65:315–22.
87. Chow VT, Tambyah PA, Yeo WM, Phoon MC, Howe J. Diagnosis of nipah virus encephalitis by electron microscopy of cerebrospinal fluid. J Clin Virol. 2000;19:143–7.
88. Itoh Y, Takahashi M, Fukuda M, Shibayama T, Ishikawa T, Tsuda F, Tanaka T, Nishizawa T, Okamoto H. Visualization of TT virus particles recovered from the sera and feces of infected humans. Biochem Biophys Res Commun. 2000;279:718–24.
89. Khaskhely NM, Uezato H, Kamiyama T, Maruno M, Kariya KI, Oshiro M, Nonaka S. Association of human papillomavirus type 6 with a verruciform xanthoma. Am J Dermatopathol. 2000;22:447–52.
90. Otsu R, Ishikawa A, Mukae K. Detection of small round structured viruses in stool specimens from outbreaks of gastroenteritis by electron microscopy and reverse transcription-polymerase chain reaction. Acta Virol. 2000;44:53–5.
91. Fenner FJ, White DO. Laboratory diagnosis of viral disease. 4th ed. New York: Academic Press; 1994.
92. Hendrickson KJ. Advances in the laboratory diagnosis of viral respiratory disease. Pediatr Infect Dis J. 2004;23:S6–10.
93. Kesson AM. Respiratory virus infections. Paediatr Respir Rev. 2007;8:240–8.

94. Drew WL, Stevens GR. How your laboratory should perform viral studies (continued): isolation and identification of commonly encountered viruses. Laboratory Med. 1980;11:14–23.
95. Fan J, Hendrickson KJ, Slavatski LL. Rapid simultaneous diagnosis of infections with respiratory syncytial viruses A and B influenza viruses A and B, and human parainfluenza viruses types 1, 2 and 3 by multiplex quantitiative reverse trancription-polymerase chain reaction-enzyme hybridization assay. Clin Infect Dis. 1998;26:1397–402.
96. Fan J, Henrickson KJ. Rapid diagnosis of human parainfluenza virus type 1 infection by quantitative reverse transcription-PCR-enzyme hybridization assay. J Clin Microbiol. 1996;34:1914–7.
97. Woo PC, Chiu SS, Seto WH, Peiris M. Cost-effectiveness of rapid diagnosis of viral respiratory tract infections in pediatric patients. J Clin Microbiol. 1997;35:1579–81.
98. Falsey AR, Formica MA, Treanor JJ, Walsh EE. Comparison of quantitative reverse transcription-PCR to viral culture for assessment of respiratory syncytial virus shedding. J Clin Microbiol. 2003;41:4160–5.
99. Falsey AR, Formica MA, Walsh EE. Diagnosis of respiratory syncytial virus infection: comparison of reverse transcription-PCR to viral culture and serology in adults with respiratory illness. J Clin Microbiol. 2002;40:817–20.
100. Heid CA, Stevens J, Livak KJ, Williams PM. Real time quantitative PCR. Genome Res. 1996;6:986–94.
101. Higuchi R, Fockler C, Dollinger G, Watson R. Kinetic PCR analysis: real-time monitoring of DNA amplification reactions. Biotechnology (N Y)1993;11:1026–30.
102. Tyagi S, Kramer FR. Molecular beacons: probes that fluoresce upon hybridization. Nat Biotechnol. 1996;14:303–8.
103. Paranhos-Baccala G, Komuruan-Pradel F, Richard N, Vernet G, Lina B, Floret D. Mixed Respiratory Virus Infections. J Clin Virol. 2008;43:407–10.
104. Lassauniere R, Kresfelder T, Venter M. A novel multiplex real0time RT-PCR assay with FRET hybridizations probes for the detection and quantitation of 13 respiratory viruses. J Virol Methods. 2010;165:254–60.
105. Bellau-Pujol S, Vabret A, Legrand L, Dina J, Gouarin S, Petitjean-Lecherbonnier J, Pozzetto B, Ginevra C, Freymuth F. Development of three multiplex RT-PCR assays for the detection of 12 respiratory RNA viruses. J Virol Methods. 2005;126:53–63.
106. Kaye M, Skidmore S, Osman H, Weinbren M, Warren R. Surveillance of respiratory virus infections in adult hospital admissions using rapid methods. Epidemiol Infect. 2006;137:792–8.
107. Lam WY, Yeung AC, Tang JW, Ip M, Chan EW, Hui M, Chan PK. Rapid multiplex nested PCR for detection of respiratory viruses. J Clin Microbiol. 2007;45:3631–40.
108. Osiowy C. Direct detection of respiratory syncytial virua, prainfluenza virus and adenovirus in clinical specimens by a multiplex reverse trancription-PCR assay. J Clin Microbiol. 1998;36:3419–154.
109. Sakthivel SK, Whitaker B, Lu X, Oliveira DBL, Stockman LJ, Kamili S, Oberste MS, Erdman DD. Comparison of fast-track diagnostics respiratory pathogens multiplex real-time RT-PCR assay with in-house singleplex assays for comprehensive detection of human respiratory viruses. J Virol Methods. 2012;185:259–66.
110. Palacios G, Briese T, Kapoor V, Jabado O, Liu Z, Venter M, Zhai J, Renwick N, Grolla A, Geisbert TW, Drosten C, Towner J, Ju J, Paweska J, Nichol ST, Swanepoel R, Feldmann H, Jahrling PB, Lipkin WI. MassTag polymerase chain reaction for differential diagnosis of viral hemorrhagic fevers. Emerg Infect Dis. 2006;12:692–5.
111. Petrik J. Microarray blood testing: Pros & cons. Biologicals. 2010;38:2–8.
112. Liu J, Ochieng C, Wiersma S, Stroher U, Towner JS, Whitmer S, Nichol ST, Moore CC, Kersh GJ, Kato C, Sexton C, Petersen J, Massung R, Hercik C, Crump JA, Kibiki G, Maro A, Mujaga B, Gratz J, Jacob ST, Banura P, Scheld WM, Juma B, Onyango CO, Montgomery JM, Houpt E, Fields B. Development of a TaqMan Array Card for acute-febrile-illness outbreak investigation and surveillance of emerging pathogens, including ebola virus. J Clin Microbiol. 2016;54:49–58.

113. Harvey JJ, Chester S, Burke SA, Ansbro M, Aden T, Gose R, Sciulli R, Bai J, DesJardin L, Benfer JL, Hall J, Smole S, Doan K, Popowich MD, St George K, Quinlan T, Halse TA, Li Z, Perez-Osorio AC, Glover WA, Russell D, Reisdorf E, Whyte Jr T, Whitaker B, Hatcher C, Srinivasan V, Tatti K, Tondella ML, Wang X, Winchell JM, Mayer LW, Jernigan D, Mawle AC. Comparative analytical evaluation of the respiratory TaqMan Array Card with real-time PCR and commercial multi-pathogen assays. J Virol Methods. 2016;228:151–7.

114. Driscoll AJ, Karron RA, Bhat N, Thumar B, Kodani M, Fields BS, Whitney CG, Levine OS, O'Brien KL, Murdoch DR. Evaluation of fast-track diagnostics and TaqMan array card real-time PCR assays for the detection of respiratory pathogens. J Microbiol Methods. 2014;107:222–6.

115. Liu J, Gratz J, Amour C, Kibiki G, Becker S, Janaki L, Verweij JJ, Taniuchi M, Sobuz SU, Haque R, Haverstick DM, Houpt ER. A laboratory-developed TaqMan Array Card for simultaneous detection of 19 enteropathogens. J Clin Microbiol. 2013;51:472–80.

116. Boeckh M, Gooley TA, Myerson D, Cunningham T, Schoch G, Bowden RA. Cytomegalovirus pp65 antigenemia-guided early treatment with ganciclovir versus ganciclovir at engraftment after allogeneic marrow transplantation: a randomized double-blind study. Blood. 1996;88:4063–71.

117. Raschilas F, Wolff M, Delatour F, Chaffaut C, De Broucker T, Chevret S, Lebon P, Canton P, Rozenberg F. Outcome of and prognostic factors for herpes simplex encephalitis in adult patients: results of a multicenter study. Clin Infect Dis. 2002;35:254–60.

118. Boonham N, Kreuze J, Winter S, van der Vlugt R, Bergervoet J, Tomlinson J, Mumford R. Methods in virus diagnostics: from ELISA to next generation sequencing. Virus Res. 2014;186:20–31.

119. Fischer N, Indenbirken D, Meyer T, Lutgehetmann M, Lellek H, Spohn M, Aepfelbacher M, Alawi M, Grundhoff A. Evaluation of unbiased next-generation sequencing of RNA (RNA-seq) as a diagnostic method in influenza virus-positive respiratory samples. J Clin Microbiol. 2015;53:2238–50.

120. Datta S, Budhauliya R, Das B, Chatterjee S, Vanlalhmuaka VV. Next-generation sequencing in clinical virology: discovery of new viruses. World J Virol. 2015;4:265–76.

121. Woo PC, Lau SK, Chu CM, Chan KH, Tsoi HW, Huang Y, Wong BH, Poon RW, Cai JJ, Luk WK, Poon LL, Wong SS, Guan Y, Peiris JS, Yuen KY. Characterization and complete genome sequence of a novel coronavirus, coronavirus HKU1, from patients with pneumonia. J Virol. 2005;79:884–95.

122. Jeffery KJ, Read SJ, Peto TE, Mayon-White RT, Bangham CR. Diagnosis of viral infections of the central nervous system: clinical interpretation of PCR results. Lancet. 1997;349: 313–7.

123. Nigrovic LE, Chiang VW. Cost analysis of enteroviral polymerase chain reaction in infants with fever and cerebrospinal fluid pleocytosis. Arch Pediatr Adolesc Med. 2000;154: 817–21.

124. Hodinka RL. 5 - Antiviral susceptibility testing using DNA-DNA hybridization. In: Wiedbrauk DL, Farkas DH, editors. Molular Methods for Virus Detection. San Diego: Academic Press; 1995. p. 103–30.

125. Erali M, Page S, Reimer LG, Hillyard DR. Human immunodeficiency virus type 1 drug resistance testing: a comparison of three sequence-based methods. J Clin Microbiol. 2001;39:2157–65.

126. Verkruyse LA, Storch GA, Devine SM, Dipersio JF, Vij R. Once daily ganciclovir as initial pre-emptive therapy delayed until threshold CMV load > or =10000 copies/ml: a safe and effective strategy for allogeneic stem cell transplant patients. Bone Marrow Transplant. 2006;37:51–6.

127. Riddler SA, Breinig MC, McKnight JL. Increased levels of circulating Epstein-Barr virus (EBV)-infected lymphocytes and decreased EBV nuclear antigen antibody responses are associated with the development of posttransplant lymphoproliferative disease in solid-organ transplant recipients. Blood. 1994;84:972–84.

128. Iloeje UH, Yang HI, Su J, Jen CL, You SL, Chen CJ, Risk Evaluation of Viral Load E, Associated Liver Disease/Cancer-In HBVSG. Predicting cirrhosis risk based on the level of circulating hepatitis B viral load. Gastroenterology. 2006;130:678–86.

129. Ohkubo K, Kato Y, Ichikawa T, Kajiya Y, Takeda Y, Higashi S, Hamasaki K, Nakao K, Nakata K, Eguchi K. Viral load is a significant prognostic factor for hepatitis B virus-associated hepatocellular carcinoma. Cancer. 2002;94:2663–8.

130. Hirsch MS, Brun-Vézinet F, D'Aquila RT, et al. Antiretroviral drug resistance testing in adult hiv-1 infection: recommendations of an international aids society–usa panel. JAMA. 2000;283:2417–26.

131. Larder B, Kemp S. Multiple mutations in HIV-1 reverse transcriptase confer high-level resistance to zidovudine (AZT)Science. 1989;246:1155–8.

132. J-f L, Linley L, Kline R, Ziebell R, Heneine W, Johnson JA. Sensitive sentinel mutation screening reveals differential underestimation of transmitted HIV drug resistance among demographic groups. AIDS. 2016;30:1439–45.

133. Petropoulos CJ, Parkin NT, Limoli KL, Lie YS, Wrin T, Huang W, Tian H, Smith D, Winslow GA, Capon DJ, Whitcomb JM. A novel phenotypic drug susceptibility assay for human immunodeficiency virus type 1. Antimicrob Agents Chemother. 2000;44:920–8.

134. Clavel F, Hance AJ. HIV drug resistance. N Engl J Med. 2004;350:1023–35.

135. Dudley DM, Chin EN, Bimber BN, Sanabani SS, Tarosso LF, Costa PR, Sauer MM, Kallas EG, O'Connor DH. Low-cost ultra-wide genotyping using Roche/454 pyrosequencing for surveillance of HIV drug resistance. PLoS One. 2012;7:e36494.

136. Inzaule SC, Ondoa P, Peter T, Mugyenyi PN, Stevens WS, de Wit TFR, Hamers RL. Affordable HIV drug-resistance testing for monitoring of antiretroviral therapy in sub-Saharan Africa. Lancet Infect Dis. 2016;16:267–75.

137. Chen X, Zou X, He J, Zheng J, Chiarella J, Kozal MJ. HIV drug resistance mutations (DRMs) detected by deep sequencing in virologic failure subjects on therapy from hunan province, China. PLoS One. 2016;11:e0149215.

138. Lowe CF, Merrick L, Harrigan PR, Mazzulli T, Sherlock CH, Ritchie G. Implementation of next-generation sequencing for hepatitis B virus resistance testing and genotyping in a clinical microbiology laboratory. J Clin Microbiol. 2016;54:127–33.

139. Zhang Q, Liao Y, Chen J, Cai B, Su Z, Ying B, Lu X, Tao C, Wang L. Corrigendum: epidemiology study of HBV genotypes and antiviral drug resistance in multi-ethnic regions from Western China. Sci Rep. 2016;6:20451.

140. Archampong TN, Boyce CL, Lartey M, Sagoe KW, Obo-Akwa A, Kenu E, Blackard JT, Kwara A. HBV genotypes and drug resistance mutations in antiretroviral treatment-naive and treatment-experienced HBV-HIV-coinfected patients. Antivir Ther. 2016;22:13–20.

141. Woods CK, Brumme CJ, Liu TF, Chui CKS, Chu AL, Wynhoven B, Hall TA, Trevino C, Shafer RW, Harrigan PR. Automating HIV drug resistance genotyping with RECall, a freely accessible sequence analysis tool. J Clin Microbiol. 2012;50:1936–42.

142. Kudesia G, Wreghitt T. Clinical and diagnostic virology. Cambridge: Cambridge University Press; 2009.

143. Korsman SNJ, van Zyl GU, Nutt L, Andersson MI, Preiser WP. Virology, an illustrated colour text. London: Elsevier; 2012.

144. World Health Organization. Manual for the laboratory diagnosis and virology surveillance of influenza. Geneva: WHO Press; 2011.

145. Wang Y, Hedman L, Perdomo MF, Elfaitouri A, Bolin-Wiener A, Kumar A, Lappalainen M, Soderlund-Venermo M, Blomberg J, Hedman K. Microsphere-based antibody assays for human parvovirus B19V, CMV and T. gondii. BMC Infect Dis. 2016;16:8.

146. Basile AJ, Horiuchi K, Panella AJ, Laven J, Kosoy O, Lanciotti RS, Venkateswaran N, Biggerstaff BJ. Multiplex microsphere immunoassays for the detection of IgM and IgG to arboviral diseases. PLoS One. 2013;8:e75670.

147. Rizwan M, Ronnberg B, Cistjakovs M, Lundkvist A, Pipkorn R, Blomberg J. Serology in the digital age: using long synthetic peptides created from nucleic acid sequences as antigens in microarrays. Microarrays (Basel). 2016;5(3). pii: E22

148. Cleton NB, Godeke GJ, Reimerink J, Beersma MF, Doorn HR, Franco L, Goeijenbier M, Jimenez-Clavero MA, Johnson BW, Niedrig M, Papa A, Sambri V, Tami A, Velasco-Salas ZI, Koopmans MP, Reusken CB. Spot the difference-development of a syndrome based protein microarray for specific serological detection of multiple flavivirus infections in travelers. PLoS Negl Trop Dis. 2015;9:e0003580.
149. Sivakumar PM, Moritsugu N, Obuse S, Isoshima T, Tashiro H, Ito Y. Novel microarrays for simultaneous serodiagnosis of multiple antiviral antibodies. PLoS One. 2013;8:e81726.
150. Ginocchio CC. Quality assurance in clinical virology. In: Specter S, Hodinka R, Young S, Wiedbrauk D, editors. Clinical virology manual. 4th ed. Washington, DC: American Society of Microbiology; 2009; p.3–17.
151. Longo MC, Berninger MS, Hartley JL. Use of uracil DNA glycosylase to control carry-over contamination in polymerase chain reactions. Gene. 1990;93:125–8.

The Microbiome in Healthy Children

7

Yvan Vandenplas and Koen Huysentruyt

Abstract

The knowledge of the importance of the interaction between the gastro-intestinal microbiome and the human being in health and disease has accumulated exponentially during recent years. Colonization of the gastro-intestinal tract during early life is critically important for a balanced development as it will determine digestive and motility maturation, metabolic, immune and brain development in early life. The optimal healthy microbiota during early life still needs further evaluation.

Many factors, environmental and patient related, determine the composition of the microbiome. Medication administered to a pregnant women, mode of delivery, mode of feeding, medication administered to the baby, will all influence the composition of the first colonization of the gastro-intestinal tract. The discovery that mother's milk contains large amounts of prebiotic oligosaccharides and small amount of probiotic bacteria had a major impact on infant formula composition. Pre- and probiotics are commonly used as supplementation in infant formula. Prebiotic oligosaccharides stimulate the growth of bifidobacteria aiming to mimic the gastrointestinal microbiota of breastfed infants. In general, results with prebiotics in therapeutic indications are disappointing. Studies suggest that probiotic supplementation may be beneficial in prevention and management of disease, e.g. reducing the risk of necrotizing enterocolitis (NEC) in preterm infants, prevention and treatment of acute gastroenteritis in infants. Although many studies show promising beneficial effects, the long-term health benefits and eventual risks of probiotic supplementation during early life are not clear. It is likely that ongoing research will result in the use of specific probiotic organisms and/or prebiotic oligosaccharides during the first 1,000 days of life, with the

Y. Vandenplas (✉) • K. Huysentruyt
Department of Pediatrics, UZ Brussel, Vrije Universiteit Brussel,
Laarbeeklaan 101, 1090 Brussels, Belgium
e-mail: Yvan.Vandenplas@uzbrussel.be

© Springer International Publishing AG 2017
R.J. Green (ed.), *Viral Infections in Children, Volume I*,
DOI 10.1007/978-3-319-54033-7_7

goal to develop a healthy microbiota from conception over birth into the first two years of life, while lowering risk of infections and inflammatory events.

7.1   Introduction

The human being should be regarded rather as a "homo bacteriens" than a *Homo sapien*". The gastro-intestinal tract of an adult human being contains over 1.5 kg bacteria, composed of more than 1000 species. The gut of an adult contains 10–100 times more bacteria than the number of cells of the same human being. Bacteria are omnipresent: they are on the skin, in the gastrointestinal (GI) tract, and in the respiratory tract.

Well known from ancient times is that bacteria can be noxious as well as beneficial. 'Plinius The Old' recommended fermented food in the treatment of infectious gastroenteritis. The discovery of penicillin in the 1940s resulted in a better understanding of how pathogenic bacteria cause disease and how antibiotics are essential to kill these bacterial pathogens and thus, cure patients. Only recently has scientific interest focused on the possible beneficial effects of bacteria.

The awareness of the importance of the interaction between bacteria and the human being, in both health and disease, has literally exploded during recent years. Colonization of the infant gut is believed to be critically important for healthy growth as it influences gut maturation, metabolic, immune and brain development in early life. In newborns and infants, the GI microbiome is essential to develop a balanced immune system. However, the question arises immediately: do we know the optimal healthy GI microbiome for the infant and child? The answer is probably an unfortunate "NO". The GI microbiota of the breastfed baby born vaginally is, in general, considered as the "healthy microbiota", but depends on the GI microbiota of the mother. The important differences between the GI microbiota development between infants born through cesarean section versus natural delivery or standard infant formula feeding versus breast feeding, are well known. Pre-term birth, cesarean section, formula feeding, antibiotic use and malnutrition have been linked to dysbiosis, which in turn is associated with several pathologies such as necrotizing enterocolitis (NEC), inflammatory bowel diseases, colic, and allergies.

The discovery that breast milk is not sterile and that even the fetus during pregnancy is colonized by bacteria, questioned previously held dogma that a fetus needs a sterile environment and, in turn, has opened new areas for research. Knowledge of the microbiome has greatly expanded over the last few years, and growing evidence suggests that aberrant bacterial communities, early in life, can lead to disease through an altered development of the immune system. Epidemiological studies have established a clear correlation between factors that disrupt the microbiota during childhood and immune and metabolic conditions. Among these disruptive factors are cesarean section, (early) antibiotic treatment, dietary supplements with pre- and probiotics, hygiene and even pets, as the most relevant [1]. Proton pump inhibitors are well established to also alter the GI microbiome [2]. The impact of antibiotic treatment during (late) pregnancy and the development of the microbiome in the newborn has been poorly studied. The development of the microbial communities during childhood will influence later health and their disturbance might lead to disease. Recent studies suggest that interventions that modify the microbial composition open the possibility of overcoming microbial imbalances during infancy as a preventive approach.

The immune system develops abnormally if the gut is not colonized by bacteria. It has become obvious that the human would not survive in a sterile environment. The "good bacteria" are needed to develop normal immune functions, GI motility, etc. The increases of many so-called "Western diseases" have clearly been related to a condition designated as "dysbiosis", an abnormal microbiota. This dysbiosis may be caused by the presence of abnormal bacteria, or an abnormal number of some strains, or a decreased diversity, as has clearly been shown in Crohn's disease. Although the relationship between dybiosis and conditions and diseases such as obesity, diabetes, inflammatory bowel disease, irritable bowel syndrome and even autism and attention deficit disorder is clearly established, causality has not yet been demonstrated. A good example is Crohn's disease where, although the dysbiosis (lack of diversity) has been demonstrated, attempts to restore the dysbiosis through the administration of probiotics or fecal transplant have failed.

Although it is obvious that the microbiome of the skin and the respiratory tract are relevant, both in number and for health, much less is known about composition, and factors that influence development.

7.2   Prenatal Factors

The stepwise microbial gut colonization process may already be initiated prenatally. The composition of the vaginal microbiota, during pregnancy, has a smaller diversity and a higher stability, and is usually dominated by Lactobacillus species. Microbes have been shown to be present in the placenta [3, 4], amniotic fluid [5], umbilical cord blood [6, 7] and meconium [8, 9]. Data suggest that intrauterine samples harbor bacterial DNA, which are, however, not necessarily cultivable bacterial cells [1]. The bacteria detected were, in many cases, common vaginal residents [10]. Microbial contact in-utero is associated with changes in a fetal intestinal innate immune gene expression profile. Fetal and placental immune physiology may be modulated by maternal dietary intervention using specific probiotics [5].

Antibiotic treatment during pregnancy is widespread in Western countries, and accounts for 80% of prescribed medications in pregnancy [4]. Bacteria are essential for normal human development and, while antibiotic treatment during pregnancy has an important role in controlling and preventing infections, it may have undesired effects regarding the maternal and feto-placental microbiome [4].

In newborns developing early sepsis, umbilical cord blood had 100% sensitivity and 94.9% specificity as compared to peripheral venous blood culture [6]. Pregnancy results in a number of changes in the vaginal microbiome, which is significantly different between pregnant and non-pregnant women [11]. The intra-uterine micro-biota is likely to be determined by the vaginal microbiota. It has been questioned if these findings in healthy pregnancies are clinically relevant, or illustrate that the techniques to detect microbes have become so powerful that they detect shortcomings from nature. We also know that the GI microbiota of the mother is influenced by medications (especially antibiotics and anti-acid medications), diet, stress and many other factors [1].

7.3   Birth and Postnatal Factors

Among the many factors that influence the development of the GI microbiota, gestational age, cesarean section and feeding are most relevant.

Postnatal factors such as antibiotic use, diet (such as breast-feeding versus formula, and introduction of solid food), genetics of the infant and environmental exposure further configure the microbiome during early life. Together with diversification, the microbiome gradually shifts toward an adult-like composition, which is usually reached by the age of 3 years.

A NEC-associated gut microbiota has been identified in meconium samples. *Clostridium perfringens* continues to be associated with NEC from the first meconium until just before NEC onset [3]. In contrast, in post-meconium, increased numbers of staphylococci were negatively associated with NEC [3].

Antibiotic use has reached widespread prevalence and are among the most commonly prescribed drugs for children [12, 13]. Postnatal antibiotics disrupt the delicate ecosystem of the neonatal microbiome [14]. The continuous use of antibiotics early in life, as well as cesarean section, have been linked to an increased risk for various conditions such as asthma, [15] (although disputed in a least one study [16]), type 2 diabetes, inflammatory bowel disease (IBD) or cow's milk protein allergy [17–19]. These correlations are not necessarily causal. Infants exposed to antibiotics in early life might experience more severe viral infections than those who are not, suggesting that impaired viral immunity increases the risk for antibiotic administration [20]. The changes that antibiotics induce in the microbiome, both in terms of composition and "time to return to baseline", depend on the type, dose of antibiotic used and body site [21].

Mouse models complement human studies and provide insights into how disruption of the microbiota can be related to disease. In mice, data suggest that antibiotics, through a disruption of the gut microbiota, induce an increase in total fat mass [22]. The antibiotic-mediated reduction in the abundance of lactobacilli and other beneficial bacteria further resulted in a decreased induction of T helper 17 (TH17) cell response in the colon [22]. Neonatal mice treated with antibiotics have also been shown to have an increased sensitization to food allergens. The same effect has been documented with proton pump inhibitors [23].

Antibiotic treatment will decrease the number of bacteria and thus impair the microbial balance among bacteria, viruses and fungi. As an example studies in mice and early human studies have documented that antibiotic exposure can lead to an increase in GI fungal abundance (*Candida albicans*), resulting in the development of airway diseases caused by allergic responses, owing to the induction of mast cells, IL-5, IL-13 and other inflammatory mediators, as well as by impairment of the antiviral immune response [24].

Feeding is another major player in the development of the GI microbiome. Breastfed infants develop a different GI microbiome than infants fed unsupplemented formula. However, these differences have become much smaller (or even undetectable) with some prebiotic supplement formulae. When an infant is given probiotic supplements, it seems logical that these probiotic strains are found in abundance in the GI microbiome. However, formula supplementation with pre- or probiotics will still not replace all the differences between breast and formula milk, including immunoglobulin A, lactoferrin, defensins, and other immunologic factors. More lactobacilli will result in in a greater acidic intestinal content and should protect from infections. More short-chain fatty acids should also have have nutritional consequences. Oligosaccharides are not absorbed but are fermented by the GI microbiota and are thus an energy source for the colonic bacteria, mainly for different strains of bifidobacteria. One of the interesting challenges to better understand is the evolution over time of different bifidobacteria in the GI microbiome. In the infant *bifidobacteria infantis* is most abundant. This species easily metabolizes oligosaccharides. However, the microbiome of the adult contains mainly *Bifidobacterium longum* which does not easily metabolize human milk oligosaccharides but more adapted to ferment complex carbohydrates. Whether artificial oligosaccharides, such as fructo- and galacto-oligosaccharides, result in the same effect as human milk oligosaccharides is a topic of ongoing research. There is a significant diversity in oligosaccharides in breastmilk. It is generally accepted that by the age of 3 years, the composition of the adult microbiome is reached [25]. However, by the age of 3 years the development of the immune system has been skewed in a certain direction. The "window of opportunity" is between conception and before the age of three.

Finally, the environment of the newborn and infant will also determine the miro- biota composition. Cleaning a baby's pacifier through an adult sucking on it has been associated with a lower risk of developing allergy [26]. Family members have a more similar oral, gut and skin microbiota than unrelated individuals [27]. Some data suggest that early exposure to specific strains may lead to life-long colonization [28], although this has not yet been confirmed in infants fed probiotic supplemented formula. The role of contact with (indoor) pets is not clear. Living on a farm or in a city environment may also determine the risk of develop of allergic disease. The socio-economic background or environmental pollution is one additional developmental factor.

Finally, host genetics will determine the development of the microbiome. Impaired bacterial regulation by the host is a potential mechanism for the pathogenesis of some metabolic diseases [29].

7.4   The Virome and Mycobiome

Commensal eukaryotic and bacterial viruses, fungi and archaea are, together with bacteria, part of the microbiome [1]. Knowledge of their role in health outcomes is very limited. The virome is the collection of eukaryotic DNA and RNA viruses and bacterial viruses (bacteriophages). In a recent study of the infant gut virome, it was found that twins had higher similarity in their virome as compared to unrelated infants. Enterovirus, parechovirus, tombamovirus and sapovirus were the most common eukaryotic viruses, and their abundances seems to depend on environmental exposure. The most abundant bacteriophages belong to the order *Caudovirales* and to the family *Microviridae* [30].

Despite its low number, the mycobiome (commensal fungal microbes) has an important role in human health. For example, overgrowth of *Candida albicans* is associated with antibiotic treatment in immunocompromised individuals. Candida, Aspergillus, Fusarium and Cryptococcus are common in oral sites of healthy individuals, whereas Malassezia spp. and *C. albicans* are found in the skin of healthy and diseased subjects [31]. The yeast *Saccharomyces boulardii* is one of the best studied probiotics [32]. Studies addressing how the infant mycobiome develops and contributes to the development of the host immune system are needed for a comprehensive understanding of the microbiome during early-life.

7.5   Probiotic Supplements

Probiotics are living microorganism that, when administered in sufficient amount, have a health benefit for the host. Prebiotics are non-digestible food ingredients that beneficially affect the host by selectively stimulating the growth and/or activity of one or a limited number of bacteria in the colon, which in turn improve host health. Synbiotics are a mixture of pre- and probiotics.

Based upon the current literature, a case can be made for the use of specific sets of probiotic organisms with the goal of promoting a healthy pregnancy and a healthy start to life, with lowered risk of infection and inflammatory events [33]. The mechanisms of specific probiotic strains administered during the perinatal period suggest that probiotic interventions in early life can be envisaged for disease prevention in both healthy infants and those at risk for the development of chronic disease. There is evidence that manipulation of the infant microbiota by using pre- or probiotics can restore the ecological balance of the microbiota and may mitigate potential negative effects on the developing immune system, when use of antibiotics cannot be avoided. As a consequence, infant formulae are increasingly supplemented with pre- and/or probiotics, despite a lack of evidence of clear clinical benefit. The effect of probiotics on obesity, allergy, infections of the GI and/or respiratory tract, and colic have been extensively studied, providing controversial results. Even the results of meta-analyses are sometimes in contradiction [34, 35]. However, only very exceptionally, results of the supplemented group are worse than the non-supplemented group [36]. Therefore, it can be stated with a high degree of confidence that results obtained with the supplementation of pre- or probiotics show either a significant benefit or no benefit. However, never is the outcome worse. Such supplementation is also considered generally devoid of adverse effects. Therefore, a common practical approach is, since pre- and probiotics are present in breastmilk, to add them to to infant formula in order to bring the composition of the GI microbiota of formula fed infants closer to that of breastfed infants.

7.6   Conclusion

A relationship exists between the composition of the microbiome in early life and the risk for later disease. The conclusive evidence that these differences, in microbiome composition, are the etiological agent for disease, and not a consequence or association, is still missing. This understanding will primarily have to come from animal studies. The ever-changing composition of the microbiome in individual conditions makes it difficult to obtain conclusive evidence. Genetic host factors make it even more complex. More data is required to better define the "optimal healthy (GI) microbiome" in the newborn. We know that the first colonization of the newborn is a major factor that will skew or orient the development of the immune system. Since so many factors determine the composition of the gut microbiota of the mother, it has been proposed that there may be benefit in developing a "super-balanced healthy microbiota" for the first colonization of the newborn. This approach would neglect the impact of environmental and genetic factors. Microbial interventions during early life might prevent or improve several diseases such as recurrent infections and allergies. Although research into the microbiome has literally exploded during recent years, the conclusion today is that very little is actually known. Hypotheses, presumptions and theoretical concepts are not yet backed up with hard clinical evidence.

Research during the next years will evaluate the relevance of the interaction between bacteria, viruses and fungi and the human being. Better understanding of this complex relationship will result in new therapeutic options.

Terminology is one of the major weaknesses today. All products containing bacteria are named probiotics. However, the vast majority of these products are uncontrolled for quality as they are commercialized as food supplements. Whilst, for some of these products, randomized prospective blinded clinical trials reveal clinical efficacy, there are no data to support the use of the majority of the probiotic food supplements. Probiotics are living microorganisms that, when consumed in adequate amount, have a health benefit for the host. The consumer is lost in the forest of ever-changing commercialized products. Both scientific and legal authorities need to find a way to protect the consumer from misleading information.

A better understanding of the composition and complex interaction of the GI microbiota will change health care in the coming years.

References

1. Tamburini S, Shen N, Wu HC, Clemente JC. The microbiome in early life: implications for health outcomes. Nat Med. 2016;22:713–22.
2. Clooney AG, Bernstein CN, Leslie WD, Vagianos K, Sargent M, Laserna-Mendieta EJ, Claesson MJ, Targownik LE. A comparison of the gut microbiome between long-term users and non-users of proton pump inhibitors. Aliment Pharmacol Ther. 2016;43:974–84.
3. Prince AL, Antony KM, Chu DM, Aagaard KM. The microbiome, parturition, and timing of birth: more questions than answers. J Reprod Immunol. 2014;104–105:12–9.
4. Kuperman AA, Koren O. Antibiotic use during pregnancy: how bad is it? BMC Med. 2016;14:91.
5. Rautava S, Collado MC, Salminen S, Isolauri E. Probiotics modulate host-microbe interaction in the placenta and fetal gut: a randomized, double-blind, placebo-controlled trial. Neonatology. 2012;102:178–84.
6. Meena J, Charles MV, Ali A, Ramakrishnan S, Gosh S, Seetha KS. Utility of cord blood culture in early onset neonatal sepsis. Australas Med J. 2015;8:263–7.
7. Jiménez E, et al. Isolation of commensal bacteria from umbilical cord blood of healthy neonates born by cesarean section. Curr Microbiol. 2005;51:270–4.
8. Gosalbes MJ, Vallès Y, Jiménez-Hernández N, Balle C, Riva P, Miravet-Verde S, de Vries LE, Llop S, Agersø Y, Sørensen SJ, Ballester F, Francino MP. High frequencies of antibiotic resistance genes in infants' meconium and early fecal samples. J Dev Orig Health Dis. 2016;7:35–44.
9. Jiménez E, Marín ML, Martín R, Odriozola JM, Olivares M, Xaus J, Fernández L, Rodríguez JM. Is meconium from healthy newborns actually sterile? Res Microbiol. 2008;159:187–93.
10. Goldenberg RL, Culhane JF, Iams JD, Romero R. Epidemiology and causes of preterm birth. Lancet. 2008;371:75–84.
11. Romero R, Hassan SS, Gajer P, Tarca AL, Fadrosh DW, Nikita L, Galuppi M, Lamont RF, Chaemsaithong P, Miranda J, Chaiworapongsa T, Ravel J. The composition and stability of the vaginal microbiota of normal pregnant women is different from that of non-pregnant women. Microbiome. 2014;2:4.
12. Hersh AL, Shapiro DJ, Pavia AT, Shah SS. Antibiotic prescribing in ambulatory pediatrics in the United States. Pediatrics. 2011;128:1053–61.
13. Hicks LA, Bartoces MG, Roberts RM, Suda KJ, Hunkler RJ, Taylor Jr TH, Schrag SJ. US outpatient antibiotic prescribing variation according to geography, patient population, and provider specialty in 2011. Clin Infect Dis. 2015;60:1308–16.
14. Arrieta MC, Stiemsma LT, Amenyogbe N, Brown EM, Finlay B. The intestinal microbiome in early life: health and disease. Front Immunol. 2014;5:427.
15. Hoskin-Parr L, Teyhan A, Blocker A, Henderson AJ. Antibiotic exposure in the first 2 years of life and development of asthma and other allergic diseases by 7.5 years: a dose-dependent relationship. Pediatr Allerg Immunol. 2013;24:762–71.
16. Örtqvist AK, Lundholm C, Kieler H, Ludvigsson JF, Fall T, Ye W, Almqvist C. Antibiotics in fetal and early life, and subsequent childhood asthma: nationwide population-based study with sibling analysis. Br Med J. 2014;349:g6979.
17. Kronman MP, Zaoutis TE, Haynes K, Feng R, Coffin SE. Antibiotic exposure and IBD development among children: a population-based cohort study. Pediatrics. 2012;130:e794–803.
18. Metsälä J, Lundqvist A, Virta LJ, Kaila M, Gissler M, Virtanen SM. Mother's and offspring's use of antibiotics, and infant allergy to cow's milk. Epidemiology. 2013;24:303–9.
19. Mikkelsen KH, Knop FK, Frost M, Hallas J, Pottegård A. Use of antibiotics and risk of type 2 diabetes: a population-based case-control study. J Clin Endocrinol Metab. 2015;100:3633–40.
20. Semic-Jusufagic A, Belgrave D, Pickles A, Telcian AG, Bakhsoliani E, Sykes A, Simpson A, Johnston SL, Custovic A. Assessing the association of early-life antibiotic prescription with asthma exacerbations, impaired antiviral immunity and genetic variants in 17q21: a population-based birth-cohort study. Lancet Respir Med. 2014;2:621–30.

21. Langdon A, Crook N, Dantas G. The effects of antibiotics on the microbiome throughout development and alternative approaches for therapeutic modulation. Genome Med. 2016;8:39.
22. Cox LM, Yamanishi S, Sohn J, Alekseyenko AV, Leung JM, Cho I, Kim SG, Li H, Gao Z, Mahana D, Zárate Rodriguez JG, Rogers AB, Robine N, Loke P, Blaser MJ. Altering the intestinal microbiota during a critical developmental window has lasting metabolic consequences. Cell. 2014;158:705–21.
23. Walker AW, Flint HJ. Editorial: further evidence that proton pump inhibitors may impact on the gut microbiota. Aliment Pharmacol Ther. 2016;43:1104–5.
24. Gonzalez-Perez G, Hicks AL, Tekieli TM, Radens CM, Williams BL, Lamousé-Smith ES. Maternal antibiotic treatment impacts development of the neonatal intestinal microbiome and antiviral immunity. J Immunol. 2016;196:3768–79.
25. Yatsunenko T, Rey FE, Manary MJ, Trehan I, Dominguez-Bello MG, Contreras M, Magris M, Hidalgo G, Baldassano RN, Anokhin AP, Heath AC, Warner B, Reeder J, Kuczynski J, Caporaso JG, Lozupone CA, Lauber C, Clemente JC, Knights D, Knight R, Gordon JI. Human gut microbiome viewed across age and geography. Nature. 2012;486:222–7.
26. Hesselmar B, Sjöberg F, Saalman R, Aberg N, Adlerberth I, Wold AE. Pacifier cleaning practices and risk of allergy development. Pediatrics. 2013;131:e1829–e37.
27. Song SJ, Lauber C, Costello EK, Lozupone CA, Humphrey G, Berg-Lyons D, Caporaso JG, Knights D, Clemente JC, Nakielny S, Gordon JI, Fierer N, Knight R. Cohabiting family members share microbiota with one another and with their dogs. Elife. 2013;2:e00458.
28. Faith JJ, Guruge JL, Charbonneau M, Subramanian S, Seedorf H, Goodman AL, Clemente JC, Knight R, Heath AC, Leibel RL, Rosenbaum M, Gordon JI. The long-term stability of the human gut microbiota. Science. 2013;341:1237439.
29. Goodrich JK, Waters JL, Poole AC, Sutter JL, Koren O, Blekhman R, Beaumont M, Van Treuren W, Knight R, Bell JT, Spector TD, Clark AG, Ley RE. Human genetics shape the gut microbiome. Cell. 2014;159:789–99.
30. Lim ES, Zhou Y, Zhao G, Bauer IK, Droit L, Ndao IM, Warner BB, Tarr PI, Wang D, Holtz LR. Early-life dynamics of the human gut virome and bacterial microbiome in infants. Nat Med. 2015;21:1228–34.
31. Gaitanis G, Magiatis P, Hantschke M, Bassukas ID, Velegraki A. The Malassezia genus in skin and systemic diseases. Clin Microbiol Rev. 2012;25:106–41.
32. Moré MI, Swidsinski A. Saccharomyces boulardii CNCM I-745 supports regeneration of the intestinal microbiota after diarrheic dysbiosis—a review. Clin Exp Gastroenterol. 2015;8: 237–55.
33. DiGiulio DB, Romero R, Amogan HP, Kusanovic JP, Bik EM, Gotsch F, Kim CJ, Erez O, Edwin S, Relman DA. Microbial prevalence, diversity and abundance in amniotic fluid during preterm labor: a molecular and culture-based investigation. PLoS One. 2008;3:e3056.
34. Pelucchi C, Chatenoud L, Turati F, Galeone C, Moja L, Bach JF, La Vecchia C. Probiotics supplementation during pregnancy or infancy for the prevention of atopic dermatitis: a meta-analysis. Epidemiology. 2012;23:402–14.
35. Kim SO, Ah YM, Yu YM, Choi KH, Shin WG, Lee JY. Effects of probiotics for the treatment of atopic dermatitis: a meta-analysis of randomized controlled trials. Ann Allergy Asthma Immunol. 2014;113:217–26.
36. Taylor AL, Dunstan JA, Prescott SL. Probiotic supplementation for the first 6 months of life fails to reduce the risk of atopic dermatitis and increases the risk of allergen sensitization in high-risk children: a randomized controlled trial. J Allergy Clin Immunol. 2007;119:184–91.

Viral-Bacterial Interactions in Childhood Respiratory Tract Infections

8

Alicia Annamalay and Peter Le Souëf

Abstract

Acute respiratory infection (ARI) is an important cause of childhood morbidity and mortality worldwide. ARIs are caused primarily by viruses and bacteria that are often co-detected in respiratory specimens. Although viral-bacterial co-infections are frequently reported in children with ARI, their clinical significance and the mechanisms leading to ARI are not well understood. The respiratory tract is a reservoir of a diverse community of microorganisms, including both commensals and potential pathogens and there is growing evidence that the interactions between viruses and bacteria play a key role in the development of ARI. A better understanding of the interactions between viruses and bacteria in the respiratory tract may enhance insight into the pathogenesis of ARI, and potentially reveal new prevention and treatment strategies. This chapter summarizes the current knowledge on viruses, bacteria and viral-bacterial interactions in childhood ARI and the possible mechanisms by which these interactions may lead to disease.

8.1 Introduction

Acute respiratory infection (ARI), both upper and lower, is a significant cause of childhood morbidity and mortality worldwide. Lower respiratory infection (LRI), including pneumonia, is among the leading causes of childhood mortality worldwide, accounting for close to a million deaths in children under 5 years of age in 2013 [1]. Viruses and bacteria can be detected in most children with ARI. However, both are also frequently detected in asymptomatic children, and hence, the clinical

A. Annamalay (✉) • P. Le Souëf
School of Paediatrics and Child Health, University of Western Australia,
Perth, WA 6840, Australia
e-mail: alicia.annamalay@uwa.edu.au

© Springer International Publishing AG 2017
R.J. Green (ed.), *Viral Infections in Children, Volume I*,
DOI 10.1007/978-3-319-54033-7_8

significance of their detection has long been debated. Furthermore, a wide range of prevalence rates are reported for the detection of viruses and bacteria, most likely due to differences in case definitions, diagnostic tools and methodologies used between studies and hence, the epidemiology and etiology of ARI are still not clear. Viruses and bacteria are often co-detected in respiratory samples from children with ARI and their interaction is likely to play a key role in the development of disease. However, the clinical significance of viral-bacterial co-infections and the mechanisms leading to ARI are not well understood.

Viruses are identified more frequently than bacteria in children with ARI [2]. Respiratory syncytial virus (RSV) has long been believed to be the most important viral cause of childhood ARI, particularly bronchiolitis, accounting for at least three million hospitalizations and up to 200,000 deaths each year [3]. Human rhinovirus (RV) is the leading cause of upper respiratory infections (URI), but is also an important cause of LRI including pneumonia and bronchiolitis, and accounts for the majority of asthma attacks in children [4, 5]. Advances in molecular methods such as polymerase chain reaction (PCR) in recent years have led to the identification of new viruses and viral species, and the importance of other viruses is now widely acknowledged. Adenovirus, influenza virus, parainfluenza virus, human metapneumovirus, coronavirus and bocavirus are among other commonly identified respiratory viruses known to contribute to the burden of childhood ARI.

Although less frequently detected than viruses, bacteria were traditionally believed to be the cause of more severe ARI-associated morbidity and mortality, particularly in developing countries [6, 7]. However, traditional diagnostic methods such as blood culture, which are still considered the gold standard in most settings, lack sensitivity and hence, the disease burden attributable to specific bacteria is not well understood. The recent advent of sequencing technologies such as 16S rRNA gene sequencing have led to the discovery of far more diverse respiratory microbial communities than previously recognized [8]. However, advanced diagnostics are largely limited to countries with the most resources and studies detecting a comprehensive range of pathogens remain scarce or non-existent. *Streptococcus pneumoniae, Haemophilus influenzae, Moraxella catarrhalis* and *Staphylococcus aureus* are acknowledged as the most common and important bacterial pathogens in childhood ARI, although more recent studies have reported substantial reductions of ARIs due to pneumococcus and *Haemophilus influenzae type B* (Hib), most likely owing to the widespread introduction of conjugate vaccines [9].

The respiratory tract is a reservoir of a diverse community of microorganisms, both commensals and potential pathogens [10]. There is growing evidence that synergistic and antagonistic interactions between viruses and bacteria play a key role in the development of ARI. A better understanding of the interactions between viruses and bacteria in the respiratory tract may provide improved insight into the pathogenesis of ARI, and potentially reveal new prevention and treatment strategies. This chapter summarizes the current knowledge on viruses, bacteria and viral-bacterial interactions in childhood ARI and the possible mechanisms by which these interactions may lead to disease.

8.2   Respiratory Viruses

Respiratory viruses are an important cause of both upper and lower ARI. The burden of viral respiratory infections is greatest in children, with infants and preschool children experiencing an estimated 6–10 viral infections annually [11]. The role of viruses in URI is well established, with 90% of URI caused by viruses compared with only 10% caused by bacteria [12]. Respiratory viruses are also commonly identified in children with LRI, and although their role in LRI is acknowledged, evidence for the clinical significance of respiratory viruses in LRI is still debated. Lung aspiration is considered the gold–standard for sampling since it is obtained directly from the site of infection and would indicate etiological significance for LRI, but is limited due to its invasive nature and rate of complications [11]. Upper respiratory tract samples such as nasopharyngeal aspirates are easily obtained and hence, more frequently used in etiological studies. However, it is argued that viruses identified in the nasopharynx represent only colonization and are not indicative of LRI. Nonetheless, viruses of the upper respiratory tract are more commonly identified in children with LRI than in asymptomatic, healthy children. A 2015 systematic review and meta–analysis of case-control studies of children with and without LRI found evidence for causal attribution of RSV, influenza, parainfluenza virus, human metapnuemovirus and RV in children presenting with LRI compared to asymptomatic or healthy children [13]. However, there was no significant difference in the detection of adenovirus, bocavirus or coronavirus between cases and controls.

While RSV and RV are widely acknowledge to be the most common and important viral causes of ARI in children, recent advances in molecular methods have highlighted the significance of other respiratory viruses.

8.2.1   Common Respiratory Viruses

8.2.1.1 Respiratory Syncytial Virus

RSV has been considered the most important viral cause of ARI in children, although its contribution is variable [14, 15]. RSV is a seasonal virus that peaks in the cold season in temperate climates and in the rainy season in tropical climates [15]. It affects about 90% of infants and young children by the age of 2 years, with peak rates occurring in infants 6 weeks to 6 months of age [15]. There are two subtypes of RSV, A and B, which circulate concurrently, although some studies have suggested that RSV-A may be more virulent than RSV-B [16–18]. In 2005, there were an estimated 33.8 million new episodes of RSV-associated ARI worldwide in children less than 5 years of age, with at least 3.4 million RSV-associated ARI hospitalizations, and an estimated 66,000–199,000 deaths, 99% of which occurred in developing countries [3]. Hence, RSV is justifiably seen to be the most important cause of childhood ARI and a major cause of ARI-associated hospital admission [3]. Although RSV has been associated with higher rates of hospitalization than other respiratory viruses in several studies, estimates of RSV-associated ARI incidence and hospitalization are highly variable between and within regions, likely due to methodological differences [15].

8.2.1.2 Human Rhinovirus

RV is the most commonly identified respiratory virus in both adults and children and was found to be responsible for approximately two-thirds of cases of the common cold [19]. Hence, it is frequently referred to as the "common cold" virus. However, there is now abundant evidence from experimental and observational studies to support the role of RV as a lower respiratory tract pathogen. Early experiment studies with RV suggested that viral replication was optimal at 33 °C (91.4 °F) and was reduced at 37 °C (98.6 °F) and 39 °C (102.2 °F) [20, 21]. However, more recent studies have shown minimal differences in replication capacities at 33 °C (91.4 °F) and 37 °C (98.6 °F) for eight different RV strains, including when viruses were cultured and titrated at the same temperature [22]. Hence, RV is now recognised as an important cause of LRI including pneumonia and bronchiolitis and importantly, accounts for the majority of asthma attacks in children [4, 5]. Several studies worldwide have reported RV as the most common virus identified in children with LRI, with identification rates of up to 63% in some populations [23].

RV was first isolated and associated with respiratory clinical disease in human in 1956 [24] and by the 1980s, one hundred and one RV-A and RV-B serotypes, known as the reference or prototype had been preserved and distributed by the American Type Culture Collection (ATCC). The retrospective discovery of the third RV species, RV-C, reported in 2006 [25, 26], led to several new investigations of the prevalence of RV. The majority of these studies in children hospitalized with ARI found that RV-C was the most prevalent RV species and was often associated with more severe illness [5, 25–31].

8.2.1.3 Adenovirus

Adenovirus is another common viral cause of ARI. Adenoviruses are classified into seven species, A to G, and different serotypes have been implicated in different clinical syndromes [32]. While the majority of adenovirus infections present as a mild URI, adenovirus is also known to cause LRI including pneumonia, bronchiolitis and bronchitis [32, 33]. Adenoviruses can also cause gastrointestinal, ophthalmologic, genitourinary and neurological infections. Adenovirus infections are most common during infancy and early childhood and is prevalent in up to 17% of children hospitalized with ARI [32]. Adenovirus is often associated episodes of recurrent wheezing, fever, hypoxia and lengthy hospitalizations [32].

8.2.1.4 Influenza Virus

Influenza virus is an important cause of ARI morbidity and mortality, with the highest burden among children less than 5 years of age and adults over 65 years of age [34, 35]. The two main subtypes of influenza virus, A and B, routinely circulate and are responsible for seasonal flu epidemics each year. In the first study to estimate the global incidence of influenza-associated ALRI in children less than 5 years of age, there were an estimated 90 million new cases of influenza episodes, 20 million cases of influenza-associated ARI and one million cases of influenza-associated severe ARI causing 28,000–111,500 deaths worldwide in 2008 [36]. In a more recent systematic analysis of the burden of influenza in paediatric respiratory hospitalizations between 1982 and 2012, influenza was associated with an estimated 870,000 hospitalizations in children less than 5 years of age annually [37].

8.2.1.5 Human Parainfluenza Virus

Human parainfluenza virus (HPIV), first discovered in the late 1950s, is an important cause of ARI in children, accounting for 2–17% of hospitalized cases [38–40]. There are four types of HPIV, HPIV 1–4. HPIV-1 to HPIV-3 are common causes of ARI in infants and young children, while HPIV-4 is less commonly associated with respiratory illness [41]. Serologic studies have demonstrated that almost all children between 6 and 10 years of age have evidence of past infection, suggesting mild or asymptomatic primary infections [42].

8.2.1.6 Human Metapnuemovirus

Human metapneumovirus (hMPV) was first isolated from young children with respiratory tract disease the Netherlands in 2001 [43]. Although hMPV was only recently discovered, phylogenetic analysis showed that hMPV has been circulating in humans for at least 50 years [44]. Clinical symptoms of hMPV are similar to those caused by RSV, ranging from URI to severe bronchiolitis and pneumonia [43, 45]. In a study evaluating the burden of hMPV infections in children in the USA, hMPV was detected in 6% of children hospitalized with ARI, 7% of children in outpatient clinics and 7% of children examined in emergency departments, with the greatest burden in children less than 1 year of age [46]. Other studies of hMPV in children hospitalized with ARI have reported prevalences as high as11–25% [47–50].

8.2.1.7 Human Coronavirus

Human coronavirus is often associated in respiratory illness. Four human coronaviruses (HCoV-229E, HCoV-OC43, HCoV-NL63 and HCoV-HKU1) are endemic in most populations and are associated with mild, self-limiting respiratory illnesses. Another two human coronaviruses, SARS-CoV and MERS-CoV cause severe respiratory syndromes and present a significant threat with their high fatality rates. The four non-severe human coronaviruses are implicated in both URI and LRI, with reported prevalence of approximately 10% in children hospitalized with ARI [51].

8.2.1.8 Human Bocavirus

Human bocavirus 1 (hBoV1) was first discovered in 2005 from patients with LRI in Sweden and was the first virus to be discovered by molecular virus screening [52]. Since then, three additional species of HboV, hBoV2, hBoV3 and hBoV4 have been discovered, [53–55] although these species are found in the gastrointestinal tract and have been associated with gastroenteritis [53–55]. In contrast, HboV1 is associated with respiratory illness, and more specifically childhood ARI, with reported prevalence ranging from 1.5 to 19% [56]. As with other respiratory viruses, diagnosis of hBoV1 infection is not possible by clinical presentation and for URI common symptoms include common cold-like symptoms and wheezing and for LRI clinical scenarios include pneumonia and bronchiolitis [57].

8.2.2 Viral Co-infections

The advent of improved molecular methods in recent years has increased the sensitivity in identifying viruses in children with ARI [58]. As a result, identification of multiple viruses in respiratory specimens from children with ARI is frequently reported [59, 60], with co-infection rates as high as 40–50% [23, 61]. However, the relationship between viral co-infections and severity of ARI in children is not conclusive. Some studies have reported increased risk of ARI hospitalization, increased length of hospital stay and worse clinical outcomes in children with viral co-infections [60, 62–65]. In contrast, recent systematic reviews and meta-analysis evaluating the relationships between respiratory viral co-infections in children have concluded that viral-coinfections were not associated with ARI severity [66–69].

To date, experimental studies of respiratory co-infections are scarce. An *in-vitro* study examining interactions between RSV and influenza virus demonstrated that growth of RSV was blocked by competitive infection with influenza A virus [70]. In the study by Shinjoh et al., RSV infection produced a higher peak viral load in single infections than in co-infections with influenza virus, if the infections were initiated at the same time [70]. However, if the influenza co-infection was initiated after the RSV infection, influenza growth was suppressed by RSV [70]. The study also demonstrated suppression of the growth of RSV by influenza A infection at the level of viral protein synthesis [70]. Indirect immunofluorescence revealed that a large proportion of infected cells synthesized both RSV and influenza A virus antigens, while scanning electron microscopy demonstrated that influenza A and RSV virions possessing surface antigens specific for each virus were selectively released from dually-infected cells [70].

In a mathematical model study investigating the dynamics of respiratory viral co-infections, Pinky et al. found that during co-infections, one virus could block another virus by being the first to infect the available host cells and that viral interference through immune response interaction was unlikely [71]. Interestingly, the study found that viral growth rate determines which virus will dominate a simultaneous infection [71]. For example, RV, the fastest-growing virus, reduced replication of the remaining viruses during a co-infection, while parainfluenza virus, the slowest-growing virus is suppressed in the presence of other viruses [71]. The authors of the study suggest that the blocking of one virus infection by the presence of another could be explained through resource competition and this finding has been supported by clinical studies of children with ARI [72, 73]. Canducci et al. found that co-infection of RSV and metapneumovirus in infants with ARI was a protective factor for length of hospital stay and hypoxia, when compared with RSV infection alone [72]. Marguet et al. also found shorter length of hospitalization in infants with RSV and RV co-infection comparing with single RSV infection [73].

Several other explanations for why identification of multiple viruses do not increase ARI severity have been suggested. One suggestion is that identification of a virus in a respiratory specimen using molecular methods may represent early detection of an infection, asymptomatic carriage, low-virulent infection or prolonged shedding [65]. Therefore, some cases of co-infection may simply be a subject of co-detection. It is also possible that the clinical outcome of viral co-infections is dependent on specific viral combinations, and this may explain the lack of consensus in the literature regarding whether viral co-infections is associated with an increased ARI severity. Many viral co-infection combinations have been reported, as reviewed by Scotta et al., with RSV and RV being the most frequently detected co-infection, although many studies do not even specify the viruses involved in viral co-infections [66].

8.3   Respiratory Bacteria

Although less frequently detected than viruses, bacteria are also acknowledged to be an important cause of ARI-associated morbidity and mortality. Traditional diagnostic methods such as blood culture, which are still considered the gold standard in most settings, lack sensitivity, and hence the disease burden attributable to specific bacteria is not well understood. Given the absence of a reference standard for the detection of bacterial pathogens, the prevalence of bacteria in ARI is likely to be higher than often reported. *S. pneumoniae, H. influenzae, M. catarrhalis* and *S. aureus* are acknowledged as the most common and important bacterial pathogens in childhood ARI, although more recent studies have reported substantial reductions of pneumococcal and Hib disease, most likely owing to the widespread introduction of conjugate vaccines [9]. Culture-independent techniques have also demonstrated that the human microbiome is far greater in extent than previously recognised [74] and that only 1% of all bacteria can be cultured using standard diagnostic methods [8]. However, advanced diagnostics are still rare in most clinical settings and studies detecting a comprehensive range of pathogens remain scarce or non-existent.

The lower respiratory tract and lungs were traditionally believed to be sterile and free from bacteria. However, it is now widely accepted that the lungs are constantly exposed to diverse communities of bacteria from the upper respiratory tract and this has been confirmed with the use of culture-independent techniques such as the 16S rRNA gene sequencing. In one of the first studies in the field, Hilty et al. challenged the dogma that the lower respiratory tract is sterile by showing that bronchial tree contains a characteristic microbial flora that differs between health and disease [75].

The human respiratory tract is home to a diverse community of both commensal and potential pathogenic bacteria that cause respiratory disease. The term "microbiome" was first proposed in 2001 by Joshua Lederberg and is used to describe this "ecological community of commensal, symbiotic and pathogenic microorganisms that literally share our body space." [76] In a balanced state, the respiratory microbiome is believed to play a beneficial role for the human host [77]. However, imbalances in the respiratory microbiome can contribute to the acquisition of new pathogens, carriage of multiple pathogens or interactions among pathogens that lead to respiratory disease. In order to cause respiratory disease, bacterial pathogens must first colonize the nasopharynx. Since the nasopharynx lies between the nose, sinuses, ears, larynx, and the lower respiratory tract, pathogens of the nasopharynx can be the source for both upper and lower respiratory tract infections and hence, plays an important role in both the development of disease and the spread of pathogens [78]. Colonization is believed to be a dynamic and complex microbial process involving acquisition and elimination of species, interactions among microbes and between microbes and the host, and interference by environmental factors [10]. Given these bacteria often co-exist in the same ecological niche; it is likely that highly evolved relationships exist between these bacteria and their interactions with each other play a critical role in the pathogenesis of disease [79]. Furthermore, it is also likely that these species interact with one another even during healthy states [10].

8.3.1 Common Respiratory Bacteria

8.3.1.1 *Streptococcus pneumoniae*

Streptococcus pneumoniae, or pneumococcus, is a Gram-positive, alpha-hemolytic (under aerobic conditions) or beta-hemolytic (under anaerobic conditions), facultative anaerobic member of the Streptococcaceae family [80]. It is responsible for a range of illnesses including pneumonia, meningitis, bacteremia, otitis media and sinusitis [81] and is the most commonly isolated organism in patients with community-acquired pneumonia [82]. *S. pneumoniae* colonizes the upper respiratory tract and is part of the normal flora of healthy individuals, particularly children. Although there are over 90 different serotypes, most cases of disease are caused by relatively few serotypes, with the ten most common serotypes accounting for 62% of invasive pneumococcal disease [83]. Since its isolation in 1881, there have been great efforts to treat and prevent *S. pneumoniae*. Antibiotic treatment for invasive pneumococcal infections typically includes broad-spectrum antibiotics until results of antibiotic sensitivity testing are available. However, emerging antibiotic resistance is a growing concern because of its potential negative impact on the outcome of patients who receive standard antibiotic therapy. Pneumococcal vaccines such as the pneumococcal conjugate vaccine or pneumococcal polysaccharide vaccine are now commonly administered to children globally.

8.3.1.2 *Haemophilus influenzae*

Haemophilus Influenzae is a Gram-negative, coccobacillary, facultative anaerobic pathogenic bacterium belonging to the Pasteurellaceae family [84]. It was first described in 1892 during an influenza pandemic and was mistakenly considered to be the cause of influenza until 1933 when the viral cause of influenza was known [85]. *H. influenzae* is commonly found in the upper and lower respiratory tract as a commensal but also causes a variety of both invasive infections, such as bacteremia, facial cellulitis, septic arthritis, and meningitis primarily in non-immune children under age 4 years of age, and respiratory tract infections such as pneumonia, acute otitis media, bronchitis, epiglottitis and sinusitis in children and adults [86]. There are six different *H. influenzae* capsular serotypes, a through f, in addition to nonencapsulated or nontypeable strains. The two most important human pathogens are the capsular serotype b strains and the nontypeable strains which are nonencapsulated. Non-typeable *H. influenzae* (NTHi) live exclusively in the pharynges of humans and are increasingly recognized as pathogens that cause both localized infections of the respiratory tract (middle ear spaces, sinuses, and bronchi) and systemic infections such as bacteremia and pneumonia [86].

8.3.1.3 *Moraxella catarrhalis*

Moraxella catarrhalis is a Gram-negative, aerobic, oxidase-positive diplococcus belonging to the Moracellaceae family that was first described in 1896 [87]. For most of the past century, *M. catarrhalis* was regarded as an upper respiratory tract commensal organism. However since the late 1970s, *M. catarrhalis* has been recognized as an important and common human respiratory tract pathogen [88]. Nasopharyngeal colonization is more prevalent among infants compared with adults, with colonization rates varying between 33 and 100% in infants from different parts of the world [89–91]. *M. catarrhalis* is also a common cause of otitis media in infants and children, causing 15–20% of acute otitis media episodes [88].

8.3.1.4 *Staphylococcus aureus*

Staphylococcus aureus, a Gram-positive coccal bacterium belonging to the Staphylococcaceae family, frequently colonizes the nasopharynx, respiratory tract and skin [92]. It is both a commensal bacterium and a human pathogen, with colonization rates of approximately 30% in the general population, although higher rates are observed in young children and the elderly [92, 93]. *S. aureus* is also a leading cause of bacteremia and infective endocarditis as well as osteoarticular, skin and soft tissue, pleuropulmonary, and device-related infections [93].

8.3.2   Bacterial Interactions

Several studies have investigated interactions between bacteria, although mostly in experimental and mathematical model studies. In an *in-vivo* study where *H. influenzae* was introduced into the nasopharynx of neonatal rats that had or had not been pre-colonized by *S. pneumoniae,* Margolis et al. reported that *H. influenzae* density increased when *S. pneumoniae* was present, suggesting synergism between these bacterial species [94]. However, when these two species were inoculated in the reverse order, inhibition was observed, indicating competition between both species [94]. Another *in-vivo* study by Lysenko et al. found that both *S. pneumoniae* and *H. influenzae* successfully colonized mice when each bacteria species was injected separately [95]. However, when *S. pneumoniae* was co-colonized with an *H. influenzae* strain, the density of *S. pneumoniae* was lower than when inoculated alone, and this was later proved to be fully dependent on complement- and neutrophil-mediated killing of pneumococci [96]. These findings were supported by a large epidemiological study of the 9-valent pneumococcal conjugate vaccine and prevalence of bacterial colonization in HIV-uninfected and HIV-infected children in South Africa that reported inverse associations between *S. pneumoniae* and *S. aureus* and between *S. aureus* and *H. influenzae* in HIV-uninfected children but not HIV-infected children [97].

Mathematical models investigating bacterial interactions have produced conflicting results. Using a multivariate random effects model for longitudinal data, Jacoby et al. found a positive association between *S. pneumoniae* and *H. influenzae* colonization among Aboriginal and non-Aboriginal children in Australia [98]. In contrast, Pettigrew et al. modeled bacterial colonization in children and reported *S. pneumoniae* colonization to be negatively associated with colonization by *H. influenzae* [99]. The study also reported negative associations between *S. pneumoniae* and *S. aureus* and between *H. influenzae* and *S. aureus*, but interestingly, when *H. influenzae* was present with *M. catarrhalis*, the odds of *S. pneumoniae* colonization increased by more than two-fold [99]. These studies suggest that interactions between bacteria are complex and may shift from negative to positive when additional bacteria species are present.

8.4   Epidemiology of Viral-Bacterial Co-infections

8.4.1   Historical Context

Viral-bacterial co-infections are frequently detected in children with respiratory illness, and there is strong evidence for enhanced ARI severity in children during co-infections compared with single infections [100]. While most studies report viral-bacterial co-infection rates ranging from 20 to 50%, rates as high as 66–77% have been observed [70–72]. The clinical significance of viral-bacterial co-infections and mechanisms that drive these interactions are not well understood. It is difficult to determine the relative importance of individual viruses and bacteria involved in co-infections since both viruses and bacteria can be carried commensally in the respiratory tract and their detection may reflect colonization, rather than infection. Furthermore, it is not possible to distinguish between primary and secondary infections in clinical studies, making it difficult to elucidate the interactions between viruses and bacteria in co-infections.

The earliest suggestion that viral infections predispose to secondary bacterial infections has been attributed to French physician R. T. H. Laennec, who observed that the prevalence of pneumonia increased following an influenza epidemic in 1803 [69]. In 1947, British epidemiologist, William Farr, coined the term "excess mortality" to describe the increase in number of deaths during the influenza season that were not caused by influenza itself, and developed the methodology used today to quantitate mortality in influenza epidemics [70]. The association between influenza and secondary bacterial infections came into particular influenza pandemics during the twentieth century, as well as subsequent observational and experimental studies, provided further evidence that viral infections predispose to secondary bacterial infections.

During the 1918 "Spanish flu" pandemic, over 50 million deaths occurred, most of which were not caused directly by influenza alone but rather as a result of secondary bacterial pneumonia [71]. The most frequently identified organisms in sputum, lung and blood samples of infected patients were *S. pneumoniae, H. influenzae, Streptococcus pyogenes* and *S. aureus*, and it was believed that the influenza virus acted synergistically with pathogenic bacteria resulting in increased incidence of disease and death [71]. These findings were supported by data from the 1957 "Asian flu" and 1968 "Hong Kong flu" pandemics showing that increased mortality was associated with increased incidence of bacterial pneumonia [72, 73]. The availability of antibiotics effective for secondary bacterial infections was believed to be a key factor for the lower number of deaths during the 1957 and 1968 pandemics compared with 1918. During the 2009 "swine flu" pandemic involving the H1N1 influenza virus, bacterial co-infection was frequently reported in fatal pneumonia cases, with *S. pneumoniae* being the most frequent bacteria identified [74, 75].

8.4.2   Clinical Evidence of Viral-Bacterial Co-infections in Children

The best and most studied example of respiratory viral-bacterial co-infections involves influenza virus, and influenza virus-bacterial co-infections has been well described in both adults and children, with clear associations with increased disease severity [101–103]. However, with the exception of outbreaks, influenza virus is a relatively infrequent viral pathogen compared to other respiratory viruses including RV and RSV.

RSV is commonly implicated in viral-bacterial co-infections with reported co-infection rates of up to 17.5–44% in RSV-infected children [104–108]. In children with severe bronchiolitis studied by Thorburn et al., bacteria were isolated from 42% of lower airway secretions from infants with RSV using culture methods [107]. *H. influenzae* and *S. aureus* were the most common bacteria identified and furthermore, the study reported that children with bacterial co-infection were at increased risk for bacterial pneumonia [107]. In serological study of children with community-acquired pneumonia, 39% of children had viral-bacterial co-infections, of which RSV and *S. pneumoniae* was the most common combination, accounting for 33% of cases [109]. Like influenza, both RSV and *S. pneumoniae* infection peaks during the winter months [110] and RSV has also been linked to seasonal increases in *S. pneumoniae* [111]. In a case-control study by Benet et al., co-infection of RSV and *S. pneumoniae* was more common in cases than in controls but co-infection of RV and *S. pneumoniae* was not different between cases and controls [112]. RSV-bacterial co-infection has also been associated with increased disease severity compared with RSV alone including longer hospital stays and more frequent admission to pediatric intensive care unit [108] and longer ventilator support [105, 107]. However, associations identified in clinical studies are often weak and only occasionally reach statistical significance. Additionally, some studies have reported bacterial co-infection rates below 2% in children with RSV [113–116]. In a study of infants hospitalized with RSV, bacterial co-infection was found in only 0.6% of children hospitalized for RSV-associated LRI [116].

RV is also commonly implicated in viral-bacterial co-infections. In a study of children with invasive pneumococcal disease by Techasaensiri et al. 34% of children had a viral co-infection, of which 25% were influenza, and 21% were RV [117]. The study reported that children with viral-coinfections were admitted to the pediatric intensive care unit more frequently and had longer hospital stays than children without viral-coinfections [117]. In a study of children with community-acquired pneumonia by Honkinen et al., viral-bacterial coinfection was identified in 66% of children, of which RV and *S. pneumoniae* was the most common combination, accounting for approximately 7% of cases [118]. Furthermore, the study reported that all cases of treatment failure had a viral-bacterial co-infection. Lauinger et al. found that among RV-infected children, bacterial co-infections, identified in 8% of children, were associated with increased admission to ICU [119].

Other respiratory viruses have also been implicated in bacterial co-infections in children, but to a lesser degree. It has been suggested that the pathogenesis of hMPV infection is strongly affected by bacterial co-infection with *S. pneumoniae*. In a study of children hospitalized with LRI in South Africa, Madhi et al. found that administration of conjugate pneumococcal vaccine reduced the incidence of hMPV infection and the incidence of clinical pneumonia in both HIV positive and negative patients [120]. These findings suggest that a significant proportion of hMPV-associated hospitalizations may be prevented by vaccination with pneumococcal conjugate vaccine. Adenovirus co-infection was identified in 21% of children with invasive pneumococcal disease [117]. Some studies have found associations between overall respiratory virus incidence and bacterial incidence, without distinguishing the specific pathogens involved [110]. While clinical studies confirm that viral-bacterial co-infections are common in children with ALRI, the absence of a control population in most studies makes it difficult to elucidate the clinical significance of co-infections.

8.5   Mechanisms for Viral-Bacterial Interactions

The historical context of viral-bacterial co-infections during influenza pandemics have led to a predominantly unidirectional view that primary viral infections increase the development of secondary bacterial infections leading to LRI. Viral-bacterial co-infections in children have been described in many clinical studies, with some associations with disease severity. However, in clinical studies, it is difficult to differentiate primary from secondary infections and to elucidate the clinical significance of co-infections. However, several *in-vivo* and *in-vitro* experimental model studies have proposed mechanisms to explain interactions between viruses and bacteria. Most mechanisms involve viral facilitation of secondary bacterial infections, for example through disruption of the respiratory epithelium or modulation of innate and adaptive immune responses to decrease bacterial clearance or increase bacterial adherence. However, there is also growing evidence that bacterial infections may promote secondary viral infections, though direct interactions, bacterial interference with antiviral immunity or by synergism or complementation by virulence factors that have similar functions.

8.5.1 Viral Promotion of Secondary Bacterial Infections

8.5.1.1 Decreased Bacterial Clearance

The respiratory epithelium is the primary site of host-pathogen encounter in the respiratory tract and the first line of defence against infection [121, 122]. The respiratory epithelium restricts bacterial attachment through mucociliary clearance and maintenance of cell-cell junctions, which restricts access to bacterial receptors [123]. There is evidence that primary viral infections disrupt the respiratory epithelium leading to decreased bacterial clearance. *In-vitro* studies have shown that cells infected with RV, RSV, adenovirus and influenza led to impairment of mucociliary function and consequent decreased clearance of bacteria including *S. pneumoniae* and *H. influenzae* [124–127].

Modulation of innate immune cells following a viral-infection is also believed to decrease bacterial clearance in the respiratory tract. Among host innate immune responses, alveolar macrophages are the major cell population in the normal airway and form the first line of defence against respiratory pathogens. A deficiency in alveolar macrophage-mediated phagocytosis following influenza has been reported in several studies. Using a murine-model, Ghoneim et al. showed that influenza infection depleted and induced cell death of alveolar macrophages leading to impaired clearance of *S. pneumoniae* [128]. Influenza and *S. pneumoniae* co-infection in mice has been also been shown to result in synergistic stimulation of type 1 interferons (IFNs) leading to impaired recruitment of macrophages and subsequently, increased bacterial colonization [129]. Another murine-model study by Jamieson et al. showed that influenza infection resulted in decreased production of inflammatory cytokines and chemokines through virus-induced glucocorticoid production, reduced recruitment of innate immune cells to the infection site and consequently, a dramatic increase in bacterial burden [130].

8.5.1.2 Increased Bacterial Adherence

Viral infections of respiratory epithelial cells can also promote bacterial adherence to host cells. In an experimental mouse model study, Hament et al. found that *S. pneumoniae* adherence to epithelial cells was enhanced by a preceding RSV infection [131] and it was later shown through *in-vitro* and *in-vivo* studies that RSV was capable of direct binding to *S. pneumoniae* [132]. Similarly, Avadhanula et al. showed that respiratory viruses including RSV, HPIV, and influenza virus enhanced adhesion of *H. influenzae* and *S. pneumoniae* to primary immortalized cell lines but only RSV and HPIV increased receptor expression for bacteria by primary bronchial epithelial cells and A549 cells [133]. Other studies have shown that RSV virions can bind directly to *S. pneumoniae and H. influenzae* acting as a direct coupling particle between bacteria and epithelial cells and thereby increasing colonization by, and enhancing, invasiveness of bacteria [124, 134]. There is also evidence that during RSV infection, viral glycoproteins at the host cell surface, act as additional receptions for bacteria adherence [132, 134]. Respiratory viruses can also increase expression of host surface proteins to which bacteria can bind [133, 135, 136]. Host cell receptors for bacterial adherence have also been found to be exposed by viral neuraminidase activities in studies of influenza virus and *S. pneumoniae* [137, 138]. There is also evidence that viral-mediated epithelia damage can lead to exposure of the basement membrane and additional receptors for bacterial adherence [139, 140].

Although the majority of evidence for enhanced bacterial adherence during viral respiratory infection comes from studies of influenza virus or RSV, there is growing evidence for the role of RV in bacterial adhesion. In an *in-vitro* study by Wang et al., nasal epithelial cells were infected with RV, and then *S. aureus*, *S. pneumoniae*, or *H. influenzae* were added to the culture [140]. Compared with RV-uninfected control cells, the adhesion of *S. aureus*, *S. pneumoniae*, and *H. influenzae* increased significantly in RV-infected nasal epithelial cells. In another *in-vitro* study on the effects of RV infection on the adherence of S. *pneumoniae* to tracheal epithelial cells, the number of *S. pneumoniae* adhering to epithelial cells increased after RV infection [141].

8.5.2   Bacterial Promotion of Secondary Viral Infections

The historical emphasis on influenza—*S. pneumoniae* co-infections in adults and the unidirectional view that viral infections increase bacterial growth may be less relevant for children. While *S. pneumoniae* carriage rates are approximately 4% in adults, carriage rates are over 50% in children [142] and up to 80% in children under 5 in developing countries [143]. This supports growing evidence that bacterial infections may promote secondary viral infections, rather than vice versa. In a large, double-blind placebo-controlled trial in infants in South Africa, the 9-valent pneumococcal conjugate vaccine was shown to prevent 31% of pneumonias associated with respiratory viruses in children in hospital, leading to the suggestion that viruses contribute to the pathogenesis of bacterial pneumonia [144]. In another study of healthy children under 2 years of age, Verkaik et al. found that higher seroconversion rates to hMPV were associated with increased nasopharyngeal carriage of *S. pneumoniae* [145]. Furthermore, well-differentiated normal human bronchial epithelial cells pre-incubated *in-vitro* with *S. pneumoniae* resulted in increased susceptibility to infection with HMPV-enhanced green fluorescent protein, suggesting that *S. pneumoniae* can modulate HMPV infection [145].

Experimental models often show an increase in influenza virus titres following a bacterial challenge. In one study, influenza viral titres in mice were shown to increase when *S. pneumoniae* was present [146]. Subsequent mathematical modelling by Smith et al. established that the influenza infection reduced the bacterial clearance ability of alveolar macrophages and that the secondary *S. pneumoniae* infection enhanced viral release from infected cells [146]. In contrast, in another mouse model study by McCullers et al., influenza infection preceding a pneumococcal challenge primed for pneumonia led to 100% mortality [147]. Interestingly, this effect was specific for viral infection preceding bacterial infection, and reversal of the order of administration led to protection from influenza and improved survival [147].

Further evidence of bacterial promotion of viral infections has been demonstrated by enhanced RSV and hMPV infection of primary epithelial cells with the addition of bacterial lipopeptides, suggesting that bacteria facilitated viral attachment to host cells [148]. *H. influenzae* has also been shown to increase airway epithelial cell ICAM-1 and TLR3 expression, leading to enhanced binding of RV and a potentiation of RV-induced chemokine release [149].

It has also been suggested that viruses might be capable of using their microbial environment to escape immune clearance, highlighting the importance of commensal microbiota in viral infections [150].

8.6 Conclusion

Although viral-bacterial co-infections are common, the clinical significance of co-infections and the mechanisms leading to ARI are yet to be fully clarified. Complex synergistic and antagonistic interactions between viruses and bacteria most likely play a key role in the development of ARI. Viruses and bacteria of the respiratory microbiome may each influence the pathogenicity and consecutive development of infections of the other. Improving knowledge of the interactions between viruses and bacteria may lead to a better understanding of the pathogenesis of ARI and eventually to new prevention and treatment strategies.

References

1. Liu L, Oza S, Hogan D, Perin J, Rudan I, Lawn JE, et al. Global, regional, and national causes of child mortality in 2000–13, with projections to inform post-2015 priorities: an updated systematic analysis. Lancet. 2015;385(9966):430–40.
2. Selwyn BJ. The epidemiology of acute respiratory tract infection in young children: comparison of findings from several developing countries. Coordinated Data Group of BOSTID Researchers. Rev Infect Dis. 1990;12(Suppl 8):870–88.
3. Nair H, Nokes DJ, Gessner BD, Dherani M, Madhi SA, Singleton RJ, et al. Global burden of acute lower respiratory infections due to respiratory syncytial virus in young children: a systematic review and meta-analysis. Lancet. 2010;375(9725):1545–55.
4. Hayden FG. Rhinovirus and the lower respiratory tract. Rev Med Virol. 2004;14(1):17–31.
5. Bizzintino JA, Lee WM, Laing IA, Vang F, Pappas T, Zhang G, et al. Association between human rhinovirus C and severity of acute asthma in children. Eur Respir J. 2011;37(5):1037–42.
6. Forgie IM, O'Neil KP, Lloyd EN, Leinonen M, Campbell H, Whittle HC, et al. Etiology of acute lower respiratory tract infections in Gambian children: I. Acute lower respiratory tract infections in infants presenting at the hospital. Pediatr Infect Dis J. 1991;10(1):33–41.
7. Forgie IM, O'Neil KP, Lloyd EN, Leinonen M, Campbell H, Whittle HC, et al. Etiology of acute lower respiratory tract infections in Gambian children: II. Acute lower respiratory tract infection in children ages one to nine years presenting at the hospital. Pediatr Infect Dis J. 1991;10(1):42–7.

8. Staley JT, Konopka A. Measurement of in situ activities of nonphotosynthetic microrganisms in aquatic and terrestrial habitats. Annu Rev Microbiol. 1985;39:321–46.

9. Jain S, Williams DJ, Arnold SR, Ampofo K, Bramley AM, Reed C, et al. Community-acquired pneumonia requiring hospitalization among U.S. Children. N Engl J Med. 2015;372(9):835–45.

10. Bosch AATM, Biesbroek G, Trzcinski K, Sanders EAM, Bogaert D. Viral and bacterial inter-actions in the upper respiratory tract. PLoS Pathog. 2013;9(1):e1003057.

11. Glezen WP, Denny FW. Epidemiology of acute lower respiratory disease in children. N Engl J Med. 1973;288(10):498–505.

12. Fahey T, Stocks N, Thomas T. Systematic review of the treatment of upper respiratory tract infection. Arch Dis Child. 1998;79(3):225–30.

13. Shi T, McLean K, Campbell H, Nair H. Aetiological role of common respiratory viruses in acute lower respiratory infections in children under five years: a systematic review and meta-analysis. J Glob Health. 2015;5(1):010408.

14. Weber MW, Mulholland EK, Greenwood BM. Respiratory syncytial virus infection in tropi-cal and developing countries. Trop Med Int Health. 1998;3(4):268–80.

15. Simoes EAF. Respiratory syncytial virus infection. Lancet. 1999;354(9181):847.

16. Jafri H, Wu X, Makari D, Henrickson KJ. Distribution of respiratory syncytial virus subtypes A and B among infants presenting to the emergency department with lower respiratory tract infection or apnea. Pediatr Infect Dis J. 2013;32(4):335–40.

17. Martinello RA, Chen MD, Weibel C, Kahn JS. Correlation between respiratory syncytial virus genotype and severity of illness. J Infect Dis. 2002;186(6):839–42.

18. Walsh EE, McConnochie KM, Long CE, Hall CB. Severity of respiratory syncytial virus infection is related to virus strain. J Infect Dis. 1997;175(4):814–20.

19. Douglas RG. Pathogenesis of rhinovirus common colds in human volunteers. Ann Otolaryngol. 1970;79:563–71.

20. Bardin PG, Johnston SL, Pattemore PK. Viruses as precipitants of asthma symptoms II. Physiology and mechanisms. Clin Exp Allergy. 1992;22(9):809–22.

21. Corne JM, Holgate ST. Mechanisms of virus induced exacerbations of asthma. Thorax. 1997;52:380–9.

22. Papadopoulos NG, Sanderson G, Hunter J, Johnston SL. Rhinoviruses replicate effectively at lower airway temperatures. J Med Virol. 1999;58(1):100–4.

23. Chidlow GR, Laing IA, Harnett GB, Greenhill AR, Phuanukoonnon S, Siba PM, et al. Respiratory viral pathogens associated with lower respiratory tract disease among young children in the highlands of Papua New Guinea. J Clin Virol. 2012;54(3):235–9.

24. Price WH. The isolation of a new virus associated with respiratory clinical disease in humans. Proc Natl Acad Sci U S A. 1956;42:892–6.

25. Lamson D, Renwick N, Kapoor V, Liu Z, Palacios G, Ju J, et al. MassTag polymerase-chain-reaction detection of respiratory pathogens, including a new rhinovirus genotype, that caused influenza- like illness in New York State during 2004-2005. J Infect Dis. 2006;194: 1398–402.

26. Arden KE, McErlean P, Nissen MD, Sloots TP, Mackay IM. Frequent detection of human rhinoviruses, paramyxoviruses, coronaviruses, and bocavirus during acute respiratory tract infections. J Med Virol. 2006;78(9):1232–40.

27. Lau SKP, Yip CCY, Lin AWC, Lee RA, So LY, Lau YL, et al. Cinical and molecular epide-miology of human rhinovirus C in children and adults in Hong Kong reveals a possible dis-tinct human rhinovirus C subgroup. J Infect Dis. 2009;200(1):1096–103.

28. Renwick N, Schweiger B, Kapoor V, Liu Z, Villari J, Bullmann R, et al. A recently identified rhinovirus genotype is associated with severe respiratory-tract infection in children in Germany. J Infect Dis. 2007;196:1745–60.

29. Miller EK, Edwards KM, Weinberg GA, Iwane MK, Griffin MR, Hall CB, et al. A novel group of rhinoviruses is associated with asthma hospitalizations. J Allergy Clin Immunol. 2009;123:98–104.

30. Miller EK, Khuri-Bulos N, Williams JV, Shehabi AA, Faouri S, Al Jundi I, et al. Human rhinovirus C associated with wheezing in hospitalised children in the Middle East. J Clin Microbiol. 2009;46(1):85–9.
31. Linsuwanon P, Payungporn S, Samransamruajkit R, Posuwan N, Makkoch J, Theanboonlers A, et al. High prevalence of human rhinovirus C infection in Thai children with acute lower respiratory tract disease. J Infect. 2009;59(2):115–21.
32. Calvo C, García-García ML, Sanchez-Dehesa R, Román C, Tabares A, Pozo F, et al. Eight year prospective study of adenoviruses infections in hospitalized children. Comparison with other respiratory viruses. PLoS One. 2015;10(7):e0132162.
33. Kunz AN, Ottolini M. The role of adenovirus in respiratory tract infections. Curr Infect Dis Rep. 2010;12(2):81–7.
34. Iskander M, Robert B, Lambert S. The burden of influenza in children. Curr Opin Infect Dis. 2007;20(3):259–63.
35. Thompson WW, Shay DK, Weintraub E, et al. Influenza-associated hospitalizations in the United States. JAMA. 2004;292(11):1333–40.
36. Nair H, Brooks WA, Katz M, Roca A, Berkley JA, Madhi SA, et al. Global burden of respiratory infections due to seasonal influenza in young children: a systematic review and meta-analysis. Lancet. 2011;378(9807):1917–30.
37. Lafond KE, Nair H, Rasooly MH, Valente F, Booy R, Rahman M, et al. Global role and burden of influenza in pediatric respiratory hospitalizations, 1982–2012: a systematic analysis. PLoS Med. 2016;13(3):e1001977.
38. Counihan M, Shay DK, Holman RC, Lowther SA, Anderson LJ. Human parainfluenza virus-associated hospitalizations among children less than five years of age in the United States. Pediatr Infect Dis J. 2001;20(7):646–53.
39. Reed G, Jewett PH, Thompson J, Tollefson S, Wright PF. Epidemiology and clinical impact of parainfluenza virus infections in otherwise healthy infants and young children <5 years old. J Infect Dis. 1997;175(4):807–13.
40. Iwane MK, Edwards KM, Szilagyi PG, Walker FJ, Griffin MR, Weinberg GA, et al. Population-based surveillance for hospitalizations associated with respiratory syncytial virus, influenza virus, and parainfluenza viruses among young children. Pediatrics. 2004;113(6): 1758–64.
41. Henrickson KJ. Parainfluenza viruses. Clin Microbiol Rev. 2003;16(2):242–64.
42. Gardner SD. The isolation of parainfluenza 4 subtypes A and B in England and serological studies of their prevalence. J Hyg. 1969;67(3):545–50.
43. van den Hoogen BG, de Jong JC, Kuiken T, Groot R, Fouchier RAM, Osterhaus ADME. A newly discovered human pneumovirus isolated from young children with respiratory tract disease. Nat Med. 2001;7:719–24.
44. de Graaf M, Osterhaus ADME, Fouchier RAM, Holmes EC. Evolutionary dynamics of human and avian metapneumoviruses. J Gen Virol. 2008;89(12):2933–42.
45. Beneri C, Ginocchio CC, Manji R, Sood S. Comparison of clinical features of pediatric respiratory syncytial virus and human metapneumovirus infections. Infect Control Hosp Epidemiol. 2015;30(12):1240–1.
46. Edwards KM, Zhu Y, Griffin MR, Weinberg GA, Hall CB, Szilagyi PG, et al. Burden of human metapneumovirus infection in young children. N Engl J Med. 2013;368(7): 633–43.
47. Madhi SA, Ludewick H, Kuwanda L, van Niekerk N, Cutland C, Klugman KP. Seasonality, incidence, and repeat human metapneumovirus lower respiratory tract infections in an area with a high orevalence of human immunodeficiency virus type-1 infection. Pediatr Infect Dis J. 2007;26(8):693–9.
48. Boivin G, De Serres G, Côté S, Gilca R, Abed Y, Rochette L, et al. Human metapneumovirus infections in hospitalized children. Emerg Infect Dis. 2003;9(6):634–40.
49. Dollner H, Risnes K, Radtke A, Nordbo SA. Outbreak of human metapneumovirus infection in Norwegian children. Pediatr Infect Dis J. 2004;25(5):436–40.

50. Esper F, Martinello RA, Boucher D, Weibel C, Ferguson D, Landry ML, et al. A 1-year experience with human metapneumovirus in children aged <5 years. J Infect Dis. 2004;189(8):1388–96.
51. Greenberg SB. Update on human rhinovirus and coronavirus infections. Semin Respir Crit Care Med. 2016;37(04):555–71.
52. Allander T, Tammi MT, Eriksson M, Bjerkner A, Tiveljung-Lindell A, Andersson B. Cloning of a human parvovirus by molecular screening of respiratory tract samples. Proc Natl Acad Sci U S A. 2005;102(36):12891–6.
53. Arthur JL, Higgins GD, Davidson GP, Givney RC, Ratcliff RM. A novel bocavirus associated with acute gastroenteritis in Australian children. PLoS Pathog. 2009;5(4):e1000391.
54. Kapoor A, Slikas E, Simmonds P, Chieochansin T, Naeem A, Shaukat S, et al. A new bocavirus species in human stool. J Infect Dis. 2009;199(2):196–200.
55. Kapoor A, Simmonds P, Slikas E, Li L, Bodhidatta L, Sethabutr O, et al. Human bocaviruses are highly diverse, dispersed, recombination prone, and prevalent in enteric infections. J Infect Dis. 2010;201(11):1633–43.
56. Allander T. Human bocavirus. J Clin Virol. 2008;41(1):29–33.
57. Berry M, Gamieldien J, Fielding B. Identification of new respiratory viruses in the new millennium. Viruses. 2015;7(3):996.
58. Gharabaghi F, Hawan A, Drews SJ, Richardson SE. Evaluation of multiple commercial molecular and conventional diagnostic assays for the detection of respiratory viruses in children. Clin Microbiol Infect. 2011;17(12):1900–6.
59. Calvo C, García-García ML, Blanco C, Vázquez MC, Frías ME, Pérez-Breña P, et al. Multiple simultaneous viral infections in infants with acute respiratory tract infections in Spain. J Clin Virol. 2008;42(3):268–72.
60. Franz A, Adams O, Willems R, Bonzel L, Neuhausen N, Schweizer-Krantz S, et al. Correlation of viral load of respiratory pathogens and co-infections with disease severity in children hospitalized for lower respiratory tract infection. J Clin Virol. 2010;48(4):239–45.
61. Nascimento MS, de Souza Ferreira AV, Rodrigues JC, Abramovici S, de LVF SF. High rate of viral identification and coinfections in infants with acute bronchiolitis. Clinics. 2010;65:1133–7.
62. Richard N, Komurian-Pradel F, Javouhey E, Perret M, Rajoharison A, Bagnaud A, et al. The impact of dual viral infection in infants admitted to pediatric intensive care unit associated with severe bronchiolitis. Pediatr Infect Dis J. 2007;27(3):213–7.
63. Cilla G, Onate E, Perez-Yarza E, Montes M, Vicente D, Perez-Trallero E. Viruses in community-acquired pneumonia in children aged less than 3 years old: high rate of viral coinfection. J Med Virol. 2008;80:1843–9.
64. Arruda E, Jones M, Escremim de Paula F, Chong D, Bugarin G, Notario G, et al. The burden of single virus and viral coinfections on severe lower respiratory tract infections among preterm infants: a prospective birth cohort study in Brazil. Pediatr Infect Dis J. 2014;33(10):998–1003.
65. Rhedin S, Lindstrand A, Rotzén-Östlund M, Tolfvenstam T, Öhrmalm L, Rinder MR, et al. Clinical utility of PCR for common viruses in acute respiratory illness. Pediatrics. 2014;133(3):e538–e45.
66. Scotta MC, Chakr VCBG, de Moura A, Becker RG, de Souza APD, Jones MH, et al. Respiratory viral coinfection and disease severity in children: A systematic review and meta-analysis. J Clin Virol. 2016;80:45–56.
67. Lim FJ, de Klerk N, Blyth CC, Fathima P, Moore HC. Systematic review and meta-analysis of respiratory viral coinfections in children. Respirology. 2016;21(4):648–55.
68. Goka EA, Vallely PJ, Mutton KJ, Klapper PE. Single and multiple respiratory virus infections and severity of respiratory disease: a systematic review. Paediatr Respir Rev. 2014;15(4):363–70.
69. Asner SA, Science ME, Tran D, Smieja M, Merglen A, Mertz D. Clinical disease severity of respiratory viral co-infection versus single viral infection: a systematic review and meta-analysis. PLoS One. 2014;9(6):e99392.

70. Shinjoh M, Omoe K, Saito N, Matsuo N, Nerome K. In vitro growth profiles of respiratory syncytial virus in the presence of influenza virus. Acta Virol. 2000;44(2):91–7.
71. Pinky L, Dobrovolny HM. Coinfections of the respiratory tract: viral competition for resources. PLoS One. 2016;11(5):e0155589.
72. Canducci F, Debiaggi M, Sampaolo M, Marinozzi MC, Berrè S, Terulla C, et al. Two-year prospective study of single infections and co-infections by respiratory syncytial virus and viruses identified recently in infants with acute respiratory disease. J Med Virol. 2008;80(4):716–23.
73. Marguet C, Lubrano M, Gueudin M, Le Roux P, Deschildre A, Forget C, et al. In very young infants severity of acute bronchiolitis depends on carried viruses. PLoS One. 2009;4(2):e4596.
74. Turnbaugh PJ, Ley RE, Hamady M, Fraser-Liggett CM, Knight R, Gordon JI. The human microbiome project. Nature. 2007;449(7164):804–10.
75. Hilty M, Burke C, Pedro H, Cardenas P, Bush A, Bossley C, et al. Disordered microbial communities in asthmatic airways. PLoS One. 2010;5(1):e8578.
76. Lederberg J, McCray A. 'Ome sweet' omics-a genealogical treasury of words. Scientist. 2001;15:8–10.
77. Blaser MJ, Falkow S. What are the consequences of the disappearing human microbiota? Nat Rev Microbiol. 2009;7(12):887–94.
78. Garcia-Rodriguez J, Fresnadillo Martinez M. Dynamics of nasopharyngeal colonization by potential respiratory pathogens. J Antimicrob Chemother. 2002;50(Suppl S2):59–73.
79. Murphy T, Bakaletz L, Smeesters P. Microbial interactions in the respiratory tract. Pediatr Infect Dis J. 2009;28(10):121–6.
80. Ryan KJ, Ray CG, editors. Sherris medical microbiology. New York: McGraw Hill; 2004.
81. Bridy-Pappas AE, Margolis MB, Center KJ, Isaacman DJ. Streptococcus pneumoniae: description of the pathogen, disease epidemiology, treatment, and prevention. Pharmacotherapy. 2005;25(9):1193–212.
82. Torres A, Blasi F, Peetermans WE, Viegi G, Welte T. The aetiology and antibiotic management of community-acquired pneumonia in adults in Europe: a literature review. Eur J Clin Microbiol Infect Dis. 2014;33(7):1065–79.
83. Kalin M. Pneumococcal serotypes and their clinical relevance. Thorax. 1998;53(3):159–62.
84. Kuhnert P, Christensen H, editors. Pasteurellaceae: biology, genomics and molecular aspects. Poole: Caister Academic Press; 2008.
85. Taubenberger JK, Hultin JV, Morens DM. Discovery and characterization of the 1918 pandemic influenza virus in historical context. Antiviral Ther. 2007;12(4 Pt B):581–91.
86. Gilsdorf JR. What the pediatrician should know about non-typeable Haemophilus influenzae. J Infect. 2015;71(Suppl 1):S10–S4.
87. Murphy TF. Branhamella catarrhalis: epidemiology, surface antigenic structure, and immune response. Microbiol Rev. 1996;60(2):267–79.
88. Goldstein EJC, Murphy TF, Parameswaran GI. Moraxella catarrhalis, a human respiratory tract pathogen. Clin Infect Dis. 2009;49(1):124–31.
89. Aniansson G, Aim B, Andersson B, Larsson P, Nylen O, Peterson H, et al. Nasopharyngeal colonization during the first year of life. J Infect Dis. 1992;165(Suppl 1):S38–42.
90. Faden H, Harabuchi Y, Hong JJ. Epidemiology of Moraxella catarrhalis in children during the first 2 years of life: relationship to otitis media. J Infect Dis. 1994;169(6):1312–7.
91. Leach AJ, Boswell JB, Asche V, Nienhuys TG, Mathews JD. Bacterial colonization of the nasopharynx predicts very early onset and persistence of otitis media in Australian aboriginal infants. Pediatr Infect Dis J. 1994;13(11):983–9.
92. Wertheim HFL, Melles DC, Vos MC, van Leeuwen W, van Belkum A, Verbrugh HA, et al. The role of nasal carriage in Staphylococcus aureus infections. Lancet Infect Dis. 2005;5(12):751–62.
93. Tong SYC, Davis JS, Eichenberger E, Holland TL, Fowler VG. Staphylococcus aureus infections: epidemiology, pathophysiology, clinical manifestations, and management. Clin Microbiol Rev. 2015;28(3):603–61.

94. Margolis E, Yates A, Levin BR. The ecology of nasal colonization of Streptococcus pneumoniae, Haemophilus influenzae and Staphylococcus aureus: the role of competition and interactions with host's immune response. BMC Microbiol. 2010;10(1):59.
95. Lysenko ES, Ratner AJ, Nelson AL, Weiser JN. The role of innate immune responses in the outcome of interspecies competition for colonization of mucosal surfaces. PLoS Pathog. 2005;1:e1.
96. Lysenko ES, Lijek RS, Brown SP, Weiser JN. Within-host competition drives selection for the capsule virulence determinant of Streptococcus pneumoniae. Curr Biol. 2010;20(13): 1222–6.
97. Madhi SA, Adrian P, Kuwanda L, Cutland C, Albrich WC, Klugman KP. Long-term effect of pneumococcal conjugate vaccine on nasopharyngeal colonization by Streptococcus pneumoniae-and associated interactions with Staphylococcus aureus and Haemophilus influenzae colonization-in HIV-Infected and HIV-uninfected children. J Infect Dis. 2007;196: 1662–6.
98. Jacoby P, Watson K, Bowman J, Taylor A, Riley TV, Smith DW, et al. Modelling the co-occurrence of Streptococcus pneumoniae with other bacterial and viral pathogens in the upper respiratory tract. Vaccine. 2007;25(13):2458–64.
99. Pettigrew MM, Gent JF, Revai K, Patel JA, Chonmaitree T. Microbial interactions during upper respiratory tract infections. Emerg Infect Dis. 2008;14(10):1584–91.
100. Brealey JC, Sly PD, Young PR, Chappell KJ. Viral bacterial co-infection of the respiratory tract during early childhood. FEMS Microbiol Lett. 2015;362(10). pii: fnv062.
101. O'Brien KL, Walters MI, Sellman J, Quinlisk P, Regnery H, Schwartz B, et al. Severe Pneumococcal pneumonia in previously healthy children: the role of preceding influenza infection. Clin Infect Dis. 2000;30(5):784–9.
102. Palacios G, Hornig M, Cisterna D, Savji N, Bussetti A, Kapoor V, et al. Streptococcus pneumoniae coinfection is correlated with the severity of H1N1 pandemic influenza. PLoS One. 2009;4(12):e8540.
103. Martén-Loeches I, Sanchez-Corral A, Diaz E, Granada RM, Zaragoza R, Villavicencio C, et al. Community-acquired respiratory coinfection in critically ill patients with pandemic 2009 influenza A(H1N1) virus. Chest. 2011;139(3):555–62.
104. Duttweiler L, Nadal D, Frey B. Pulmonary and systemic bacterial co-infections in severe RSV bronchiolitis. Arch Dis Child. 2004;89(12):1155–7.
105. Kneyber MCJ, van Oud-Alblas HB, van Vliet M, Uiterwaal CSPM, Kimpen JLL, van Vught AJ. Concurrent bacterial infection and prolonged mechanical ventilation in infants with respiratory syncytial virus lower respiratory tract disease. Intensive Care Med. 2005;31(5): 680–5.
106. Randolph AG, Reder L, Englund JA. Risk of bacterial infection in previously healthy respiratory syncytial virus-infected young children admitted to the intensive care unit. Pediatr Infect Dis J. 2004;23(11):990–4.
107. Thorburn K, Harigopal S, Reddy V, Taylor N, van Saene HKF. High incidence of pulmonary bacterial co-infection in children with severe respiratory syncytial virus (RSV) bronchiolitis. Thorax. 2006;61(7):611–5.
108. Resch B, Gusenleitner W, Mueller WD. Risk of concurrent bacterial infection in preterm infants hospitalized due to respiratory syncytial virus infection. Acta Paediatr. 2007;96(4): 495–8.
109. Heiskanen-Kosma T, Korppi M, Jokinen C, Kurki S, Heiskanen L, Juvonen H, et al. Etiology of childhood pneumonia: serologic results of a prospective, population-based study. Pediatr Infect Dis J. 1998;17(11):986–91.
110. Kim PE, Musher DM, Glezen WP, Barradas MCR, Nahm WK, Wright CE. Association of invasive pneumococcal disease with season, atmospheric conditions, air pollution, and the isolation of respiratory viruses. Clin Infect Dis. 1996;22(1):100–6.
111. Weinberger DM, Givon-Lavi N, Shemer-Avni Y, Bar-Ziv J, Alonso WJ, Greenberg D, et al. Influence of pneumococcal vaccines and respiratory syncytial virus on alveolar pneumonia. Israel Emerg Infect Dis. 2013;19(7):1084–91.

112. Bénet T, Sylla M, Messaoudi M, Sánchez Picot V, Telles J-N, Diakite A-A, et al. Etiology and factors associated with pneumonia in children under 5 years of age in Mali: a prospective case-control study. PLoS One. 2015;10(12):e0145447.
113. Purcell K, Fergie J. Concurrent serious bacterial infections in 2396 infants and children hospitalized with respiratory syncytial virus lower respiratory tract infections. JAMA Pediatr. 2002;156(4):322–4.
114. Titus MO, Wright SW. Prevalence of serious bacterial infections in febrile infants with respiratory syncytial virus infection. Pediatrics. 2003;112(2):282–4.
115. Bloomfield P, Dalton D, Karleka A, Kesson A, Duncan G, Isaacs D. Bacteraemia and antibiotic use in respiratory syncytial virus infections. Arch Dis Child. 2004;89(4):363–7.
116. Hall CB, Powell KR, Schnabel KC, Gala CL, Pincus PH. Risk of secondary bacterial infection in infants hospitalized with respiratory syncytial viral infection. J Pediatr. 1988;113(2): 266–71.
117. Techasaensiri B, Techasaensiri C, Mejias A, McCracken GH, Ramilo O. Viral coinfections in children with invasive pneumococcal disease. Pediatr Infect Dis J. 2010;29:519–23.
118. Honkinen M, Lahti E, Österback R, Ruuskanen O, Waris M. Viruses and bacteria in sputum samples of children with community-acquired pneumonia. Clin Microbiol Infect. 2012;18(3): 300–7.
119. Lauinger IL, Bible JM, Halligan EP, Bangalore H, Tosas O, Aarons EJ, et al. Patient characteristics and severity of human rhinovirus infections in children. J Clin Virol. 2013;58(1): 216–20.
120. Madhi SA, Ludewick H, Kuwanda L, Niekerk NV, Cutland C, Little T, et al. Pneumococcal coinfection with human metapneumovirus. J Infect Dis. 2006;193(9):1236–43.
121. Proud D, Leigh R. Epithelial cells and airway diseases. Immunol Rev. 2011;242(1): 186–204.
122. Vareille M, Kieninger E, Edwards MR, Regamey N. The airway epithelium: soldier in the fight against respiratory viruses. Clin Microbiol Rev. 2011;24(1):210–29.
123. Melvin JA, Bomberger JM. Compromised defenses: exploitation of epithelial responses during viral-bacterial co-infection of the respiratory Tract. PLoS Pathog. 2016;12(9):e1005797.
124. Smith CM, Sandrini S, Datta S, Freestone P, Shafeeq S, Radhakrishnan P, et al. Respiratory syncytial virus increases the virulence of Streptococcus pneumoniae by binding to penicillin binding protein 1a. A new paradigm in respiratory infection. Am J Respir Crit Care. 2014;190:196–207.
125. Pittet LA, Hall-Stoodley L, Rutkowski MR, Harmsen AG. Influenza virus infection decreases tracheal mucociliary velocity and clearance of Streptococcus pneumoniae. Am J Respir Cell Mol Biol. 2010;42(4):450–60.
126. Suzuki K, Bakaletz LO. Synergistic effect of adenovirus type 1 and nontypeable Haemophilus influenzae in a chinchilla model of experimental otitis media. Infect Immunol. 1994;62(5): 1710–8.
127. Sajjan U, Wang Q, Zhao Y, Gruenert DC, Hershenson MB. Rhinovirus disrupts the barrier function of polarized airway epithelial cells. Am J Respir Crit Care. 2008;178(12):1271–81.
128. Ghoneim HE, Thomas PG, McCullers JA. Depletion of alveolar macrophages during influenza infection facilitates bacterial superinfections. J Immunol. 2013;191(3):1250–9.
129. Nakamura S, Davis KM, Weiser JN. Synergistic stimulation of type I interferons during influenza virus coinfection promotes Streptococcus pneumoniae colonization in mice. J Clin Invest. 2011;121(9):3657–65.
130. Jamieson AM, Yu S, Annicelli CH, Medzhitov R. Influenza virus-induced glucocorticoids compromise innate host defense against a secondary bacterial infection. Cell Host Microbe. 2010;7(2):103–14.
131. Hament JM, Aerts PC, Fleer A, Dijk H, Harmsen T, Kimpen JL, et al. Enhanced adherence of Streptococcus pneumoniae to human epithelial cells infected with respiratory syncytial virus. Pediatr Res. 2004;55:972–8.
132. Hament JM, Aerts PC, Fleer A, van Dijk H, Harmsen T, Kimpen JL, et al. Direct binding of respiratory syncytial virus to pneumococci: a phenomenon that enhances both pneumococcal

adherence to human epithelial cells and pneumococcal invasiveness in a murine model. Pediatr Res. 2005;58:1198–203.

133. Avadhanula V, Rodriguez CA, Devincenzo JP, Wang Y, Webby RJ, Ulett GC, et al. Respiratory viruses augment the adhesion of bacterial pathogens to respiratory epithelium in a viral species- and cell type-dependent manner. J Virol. 2006;80:1629–36.

134. Avadhanula V, Wang Y, Portner A, Adderson E. Nontypeable Haemophilus influenzae and Streptococcus pneumoniae bind respiratory syncytial virus glycoprotein. J Med Microbiol. 2007;56(9):1133–7.

135. Cundell DR, Gerard NP, Gerard C, Idanpaan-Heikkila I, Tuomanen EI. Streptococcus pneumoniae anchor to activated human cells by the receptor for platelet-activating factor. Nature. 1995;377(6548):435–8.

136. Kimaro Mlacha SZ, Peret TCT, Kumar N, Romero-Steiner S, Dunning Hotopp JC, Ishmael N, et al. Transcriptional adaptation of pneumococci and human pharyngeal cells in the presence of a virus infection. BMC Genomics. 2013;14(1):378.

137. McCullers JA, Bartmess KC. Role of neuraminidase in lethal synergism between influenza virus and Streptococcus pneumoniae. J Infect Dis. 2003;187(6):1000–9.

138. Huber VC, Peltola V, Iverson AR, McCullers JA. Contribution of vaccine-induced immunity toward either the HA or the NA component of influenza viruses limits secondary bacterial complications. J Virol. 2010;84(8):4105–8.

139. Plotkowski MC, Puchelle E, Beck G, Jacquot J, Hannoun C. Adherence of type I Streptococcus pneumoniae to tracheal epithelium of mice infected with influenza A/PR8 virus. Am Rev Respir Dis. 1986;134:1040–9.

140. Wang JH, Kwon HJ, Jang YJ. Rhinovirus enhances various bacterial adhesions to nasal epithelial cells simultaneously. Laryngoscope. 2009;119:1406–11.

141. Ishizuka S, Yamaya M, Suzuki T, Takahashi H, Ida S, Sasaki T, et al. Effects of rhinovirus infection on the adherence of Streptococcus pneumoniae to cultured human airway epithelial cells. J Infect Dis. 2003;188(12):1928–39.

142. Regev-Yochay G, Raz M, Dagan R, Porat N, Shainberg B, Pinco E, et al. Nasopharyngeal carriage of Streptococcus pneumoniae by adults and children in community and family settings. Clin Infect Dis. 2004;38(5):632–9.

143. Adegbola RA, DeAntonio R, Hill PC, Roca A, Usuf E, Hoet B, et al. Carriage of Streptococcus pneumoniae and other respiratory bacterial pathogens in low and lower-middle income countries: a systematic review and meta-analysis. PLoS One. 2014;9(8):e103293.

144. Madhi SA, Klugman KP. A role for Streptococcus pneumoniae in virus-associated pneumonia. Nat Med. 2004;10:811–3.

145. Verkaik NJ, Nguyen DT, de Vogel CP, Moll HA, Verbrugh HA, Jaddoe VWV, et al. Streptococcus pneumoniae exposure is associated with human metapneumovirus seroconversion and increased susceptibility to in vitro HMPV infection. Clin Microbiol Infect. 2011;17(12):1840–4.

146. Smith AM, Adler FR, Ribeiro RM, Gutenkunst RN, McAuley JL, McCullers JA, et al. Kinetics of coinfection with influenza A virus and Streptococcus pneumoniae. PLoS Pathog. 2013;9(3):e1003238.

147. McCullers JA, Rehg JE. Lethal synergism between influenza virus and Streptococcus pneumoniae: characterization of a mouse model and the role of platelet-activating factor receptor. J Infect Dis. 2002;186(3):341–50.

148. Nguyen DT, de Witte L, Ludlow M, Yüksel S, Wiesmüller K-H, Geijtenbeek TBH, et al. The synthetic bacterial lipopeptide Pam3CSK4 modulates respiratory syncytial virus infection independent of TLR activation. PLoS Pathog. 2010;6(8):e1001049.

149. Sajjan US, Jia Y, Newcomb DC, Bentley JK, Lukacs NW, LiPuma JJ, et al. H. influenzae potentiates airway epithelial cell responses to rhinovirus by increasing ICAM-1 and TLR3 expression. FASEB J. 2006;20:2121–3.

150. Kane M, Case LK, Kopaskie K, Kozlova A, MacDearmid C, Chervonsky AV, et al. Successful transmission of a retrovirus depends on the commensal microbiota. Science. 2011;334(6053):245–9.

Index

CPI Antony Rowe
Chippenham, UK
2017-10-24 21:45